Homebuilt Broomhandle Pistol

by

Gary F. Hartman

Published by
Gary F. Hartman
Lebanon, OR

Homebuilt
Broomhandle
Pistol
==============

Copyright © 2020
by Gary F. Hartman

All rights reserved. No part of this
book may be reproduced or transmitted
in any form, or by any means
(except for the inclusion of brief
quotations in reviews) without the prior
written permission of the author.

ISBN 978-0-9815399-6-6

Library of Congress
Control Number: 2020906389

Printed in U.S.A. By Ingram Content

Foreword

This book describes construction of a home made pistol that slightly resembles the old Broomhandle Mauser, Model C-96, produced from 1896-1937.

The construction of this firearm uses garage tools, to show a simple firearm can be made using only files, a drill press, bench grinder, and angle grinder with a cutoff disk, drill, a Dremel type tool, and a wire welder.

The start is buying a used gun barrel from a gun show or gun shop; these are always available from many sources.

The original Mauser pistol was semi-auto and was copied by various countries in a mix of pistol calibers. I believe a switchable version was also capable of full auto.

The gun in this book is not a semi-auto pistol, it can be built as a repeater version or a double action mechanism. And it is .22 caliber. It is a similar but simpler build to my book, "Homebuilt Firearms" 2010.

I believe the right to self defense is an inherent God given right, no government, no official, no politician or dictator can take that right away, unless the foolish allow it. And the best means of self defense is a firearm. That is accepted as the common sense rule by all free people. For that reason, this book is something to take to heart, you can learn and you can do it!

The build described here is a functional multi-shot pistol; it is something a person with the desire can make in their garage with only typical garage workbench tools. By taking time to work carefully and accurately, the pistol will function correctly.

The care and time involved will result in a usable firearm,

functional not as a professional, finely built firearm, but still useful as a simple, expedient working gun.

If you are a person who likes "do it yourself" projects, to tinker and build and who does not expect an instant result from minimal effort, like many of the newer generation, then this is a book for you.

Safety Concerns

As in any project using power tools, or sharp hand tools, safety is the main concern as you work. Always wear safety goggles, a dust mask and ear protection when using those tools requiring these items. **Do NOT cut corners on safety!**

Work safely, plan ahead and have fun!

Gary F. Hartman
Lebanon, OR
<u>www.jgenasplace.com</u>

Previous Books

Kids' Book of Adventure Projects	2008
Homebuilt Firearms	2010
Homebuilt Clocks	2011
Homebuilt .45 ACP Carbine	2013
Homebuilt Toys	2015

Table of Contents

Chapter

1. General Overview — 1
2. Barrel Preparation — 19
3. Receiver and Fr. Magazine Guide — 25
4. The Bar Magazine — 36
5. Finishing Fr. And Rear Magazine Guide — 44
6. Making the Internal Parts — 53
7. My Hammer and Sear — 72
8. Making the Positioner — 81
9. Heat Treating of Parts — 88
10. Barrel and Hand Grip — 92
11. Finishing Receiver and Final Notes — 102

Chapter 1

General Overview

I do not pretend this is some sort of fabulous design; this pistol is just a simple way to make a multiple shot firearm using garage tools.

Also anyone would agree that using fine machine shop equipment to fabricate the pieces is superior to making them by filing and grinding in a bench environment. A manufacturing enterprise could make a design using aluminum and other metals to make a very fine version beyond the scope of this book.

My purpose was simply to do this project to determine it could be done with garage tools: building a functioning firearm for anyone willing to put in the effort to do the project.

My welding is not beautiful, many parts may look crude. But, until you make and try to weld a tiny part that might have a dimension of a fraction of an inch, you have no idea of the difficulty. And your work is your own.

Tools Required

Several garage tools come in handy for the gun construction:

- Angle Grinder with cutoff disc, Figure 1.

- Bench Grinder, Figure 2.

- Files, Flat, Round types. And Hacksaw.

- A Dremel Moto Tool, Figure 3.

- Safety Glasses, Dust Mask and Ear Protectors.

- Wire Welder, Figure 4. Harbor Freight or similar.

- Drill Press with X-Y Table, Figure 5. The X-Y table allows doing precision drilling, measuring to thousandths for accurate hole placement.

- A Screw Machine #1 drill bit for the bar magazine holes for the .22 cartridges. This bit should be purchased from an industrial machine supply company and is the shorter length for machining. It is far more solid and rigid than a normal length drill bit. Also get a 1/4" double ended center drill bit. See Figure 6.

These items will give a good range of fabrication ability to cut, drill pieces, and true up the metal parts and precision drill the magazine for the pistol.

In every operation, think it through beforehand to identify the order of your steps in fabrication, and also think of possible problems, and identify what safety equipment you will need. Always keep safety as your high priority, good safety equipment can protect you from injury.

Be Safe!!

Fig. 1. Angle Grinder with Cutoff Disc.

Fig. 2. Bench Grinder.

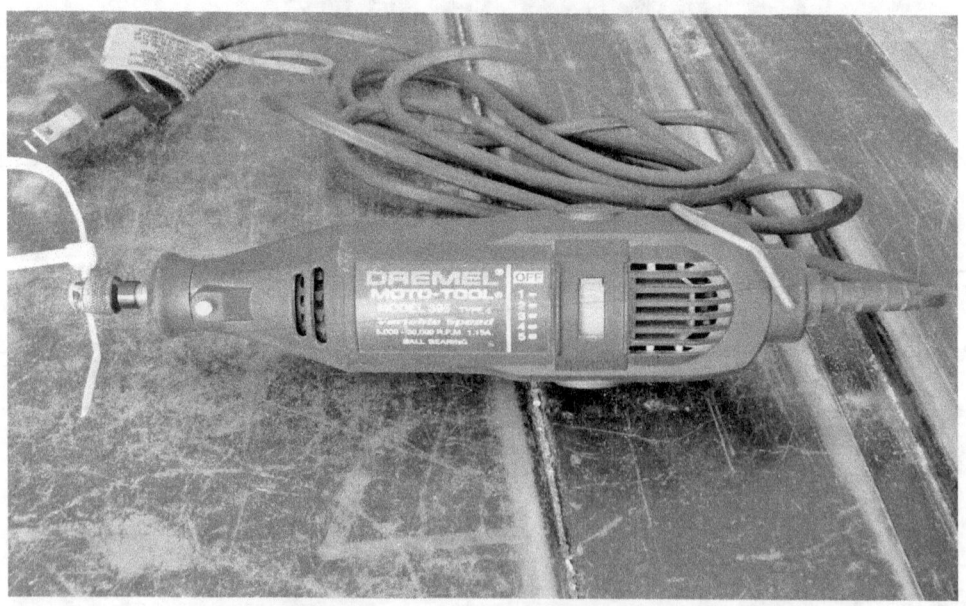

Fig. 3. Dremel Moto Tool.

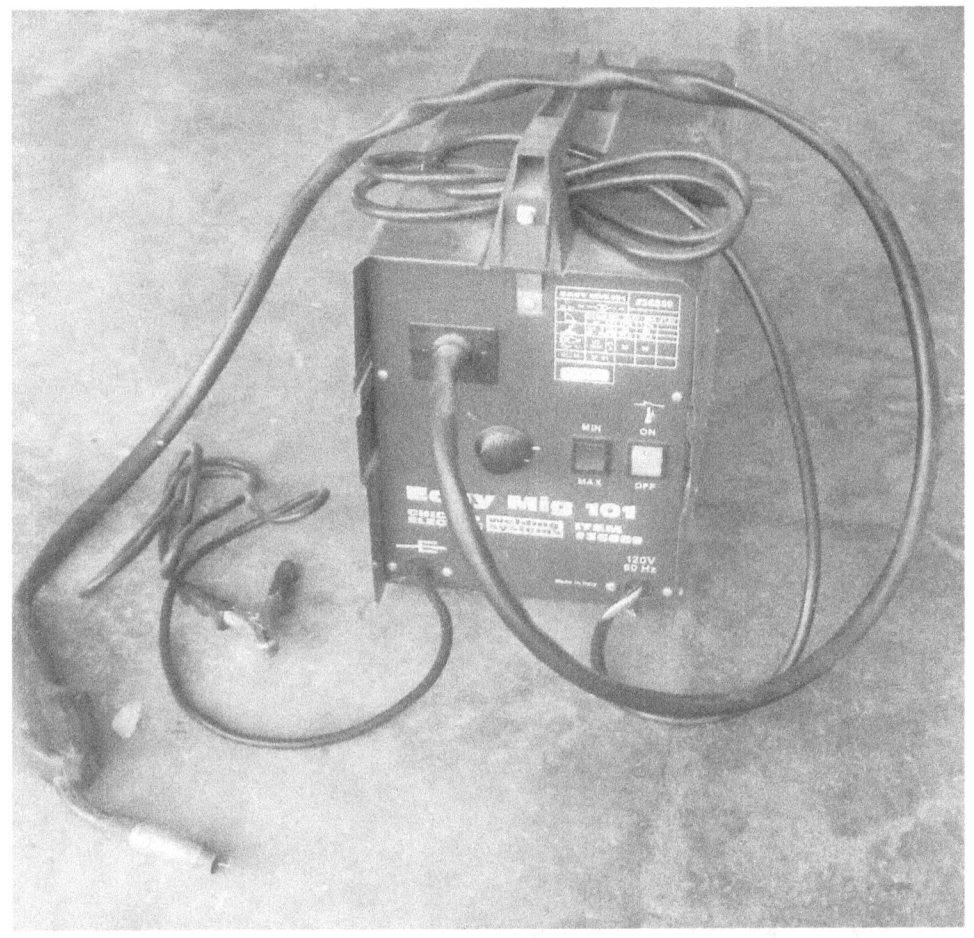

Fig. 4. Wire Welder.

The wire welder uses a motor to feed the weld wire at a speed to approximate the rate it melts and forms the weld in the metal. So, with some practice, a fabricator can adjust the motor speed to fine tune and get a reasonably smooth weld.

However, sometimes, because of the tiny parts being welded, it is difficult to always do a beautiful weld job-- at least for me.

Also the welding spits particles of molten metal which can make the job look pretty rough. The main thing is achieving a good bond between pieces you weld.

Fig. 5. Drill Press with X-Y Table.

The drill press with X-Y cross-feed table allows precision in drilling the holes in the magazine, a very important item.

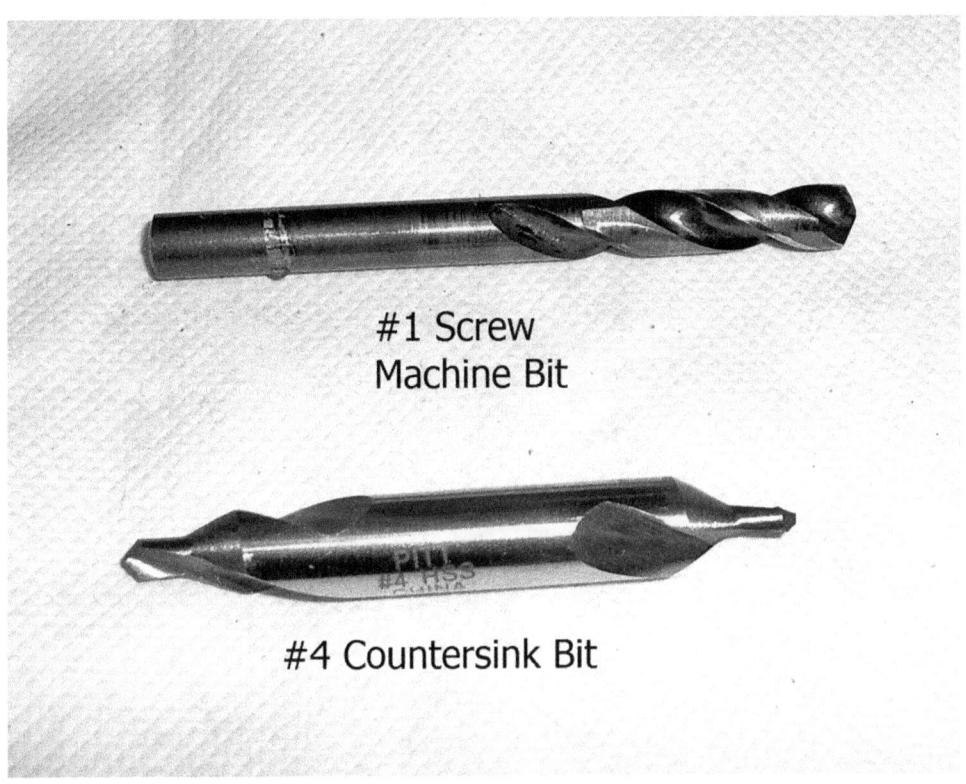

Fig. 6. #1 Screw Machine Bit and #4 Center Drill Bit.

Pistol Description

- This pistol uses similar items to a Single Action or Double Action revolver.
- A revolver has a revolving cylinder which carries the rounds, but this pistol has a rectangular steel magazine; the magazine itself ratchets through the gun and exits the gun when all rounds are fired.
- See Figure 7, which shows a bar magazine version, rather than a cylinder.

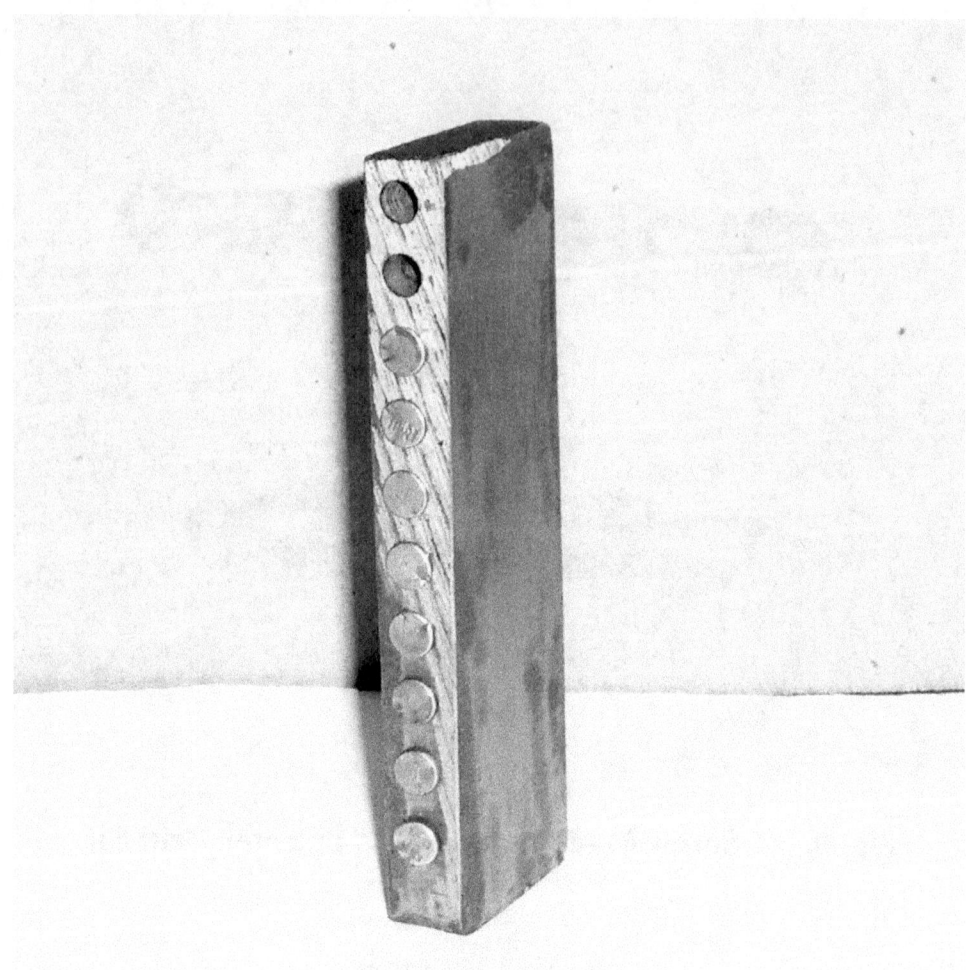

Fig. 7. A Bar Magazine.

- Next, note the hammer shaped like a "T" that incorporates a positioning rod arm which forces the magazine to advance one shot at a time as the hammer cocks.

- The versions of the pistol here discussed, is a Single Action version (requires cocking before each shot, released by the trigger.)

- Another type would be a Double Action version, where pulling the trigger lifts the hammer at the same time the magazine advances, and as the new round seats, the hammer releases from the trigger mechanism and fires the round.

The pistol appears odd because the barrel is near the bottom of the receiver.

Yet, think about this: The barrel lying almost in-line with the hand means less upward rotation due to recoil like in a normal barrel-at-the-top scenario. This is a good thing, even though this is only a .22 caliber firearm. The recoil is more directly in-line and opposed by the gripping hand. It allows a more controlled firing of the pistol.

Figure 8, on the next page, shows a simplified side view of the barrel, internal parts, front and rear guides and bar magazine set up so you can visualize the general idea before the beginning of the construction. This is only a general view to show you the function and look that the construction will take as the project proceeds.

Actual pieces are in Figure 9 with the front guide un-drilled or fitted for the two holes needed, just to give a visualization.

- a hole is required for the round to fire through as it enters the barrel.

- Also a hole to hold a 5/16" steel ball which is used as an index device to grab each hole in the bar magazine locking the bar in correct position for each shot.

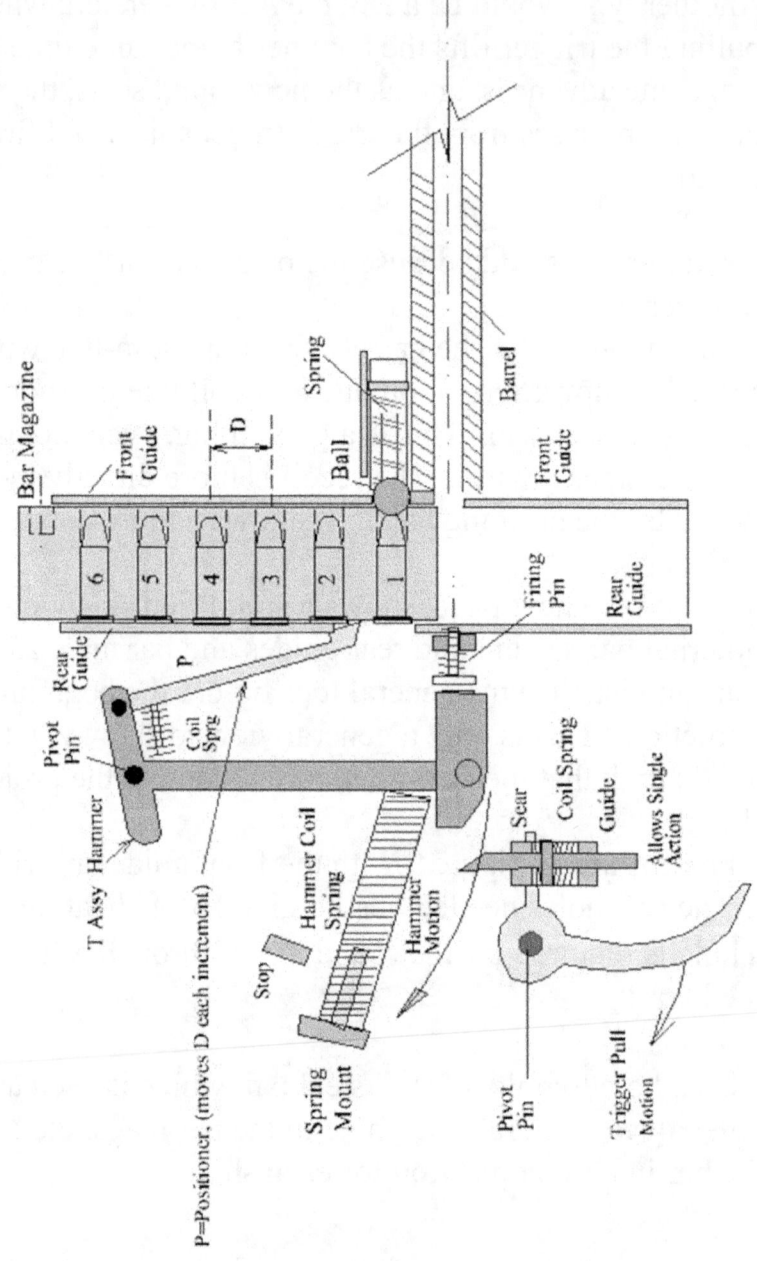

Fig. 8. Side View Details.

Fig. 9. General Layout of Initial Parts.

The main parts in front of the forward magazine guide area will be the barrel, and the locking index ball and spring or some other method to lock the bar magazine in alignment once it is in the firing position.

I originally envisioned this spring loaded ball as a method to lock the bar magazine with the round just below in alignment with the barrel. Other locking methods can be used, a spring-loaded lug, any other method which firmly locks the bar magazine at the correct location. You can envision the general look from this photo.

There will be a rear guide of perhaps 12 gauge steel similar to the front guide. Where the cartridge is to be fired, a thick steel block backing plate will be mounted, with a hole for the firing pin. This heavier piece, welded on, will withstand the rear force of the cartridge as it fires.

Above this will be the sheet steel rear guide similar to the front guide, as it does not need to resist any firing force. It will have a slot parallel to the receiver plate, centered in the guide.

- This slot will allow a "positioner" piece attached to the hammer mechanism to push each round downward as the hammer cocks, thus pushing each round to the spot where the spring loaded index ball will jump into the front hole of the magazine bar at the round just above the one to be fired and lock it into position. This insures the round aligned with the barrel will be locked in place ready for the next shot.

- Look at the series of *side view* pictures in the next few pages, which illustrate the operation of the Single Action operation of this version of the pistol.

- This is a simplified side view of the parts for the simple Single Action mechanism. The colored parts give you the shape clearly, even though the actual real part may be a bit more involved with a pivoting piece or other part attached.

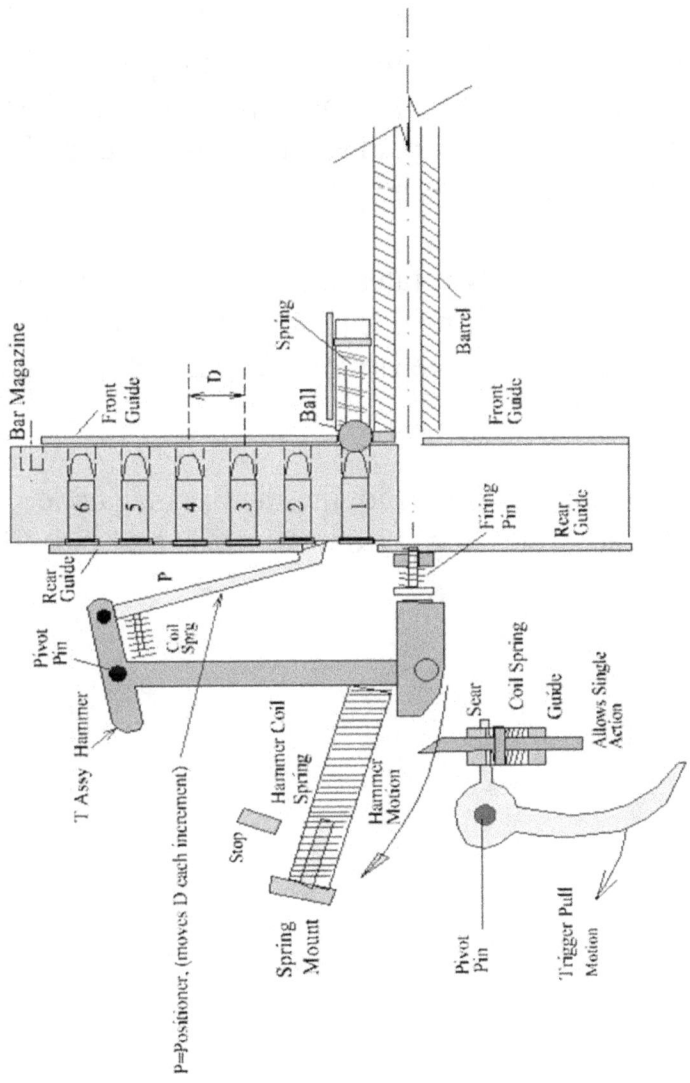

Fig. 10. Initial Position, Magazine inserted, Side View.

Figure 10 shows the mechanism at the beginning of use.

- The bar magazine has just been loaded down into its slot until the first round's front exit hole is grabbed by the index ball.

- Note the pivot pin at the top of the one piece T assembly hammer, and imagine the lower portion of the hammer swinging clockwise in an arc as the hammer spring begins to compress.

- As the T assembly rotates around the top pivot pin, in a cocking motion, it begins pushing the positioner rod, P, downward against the rim of the first cartridge in the magazine bar.

- The positioner begins moving the bar magazine downward, releasing it from the index ball. See Figure 11.

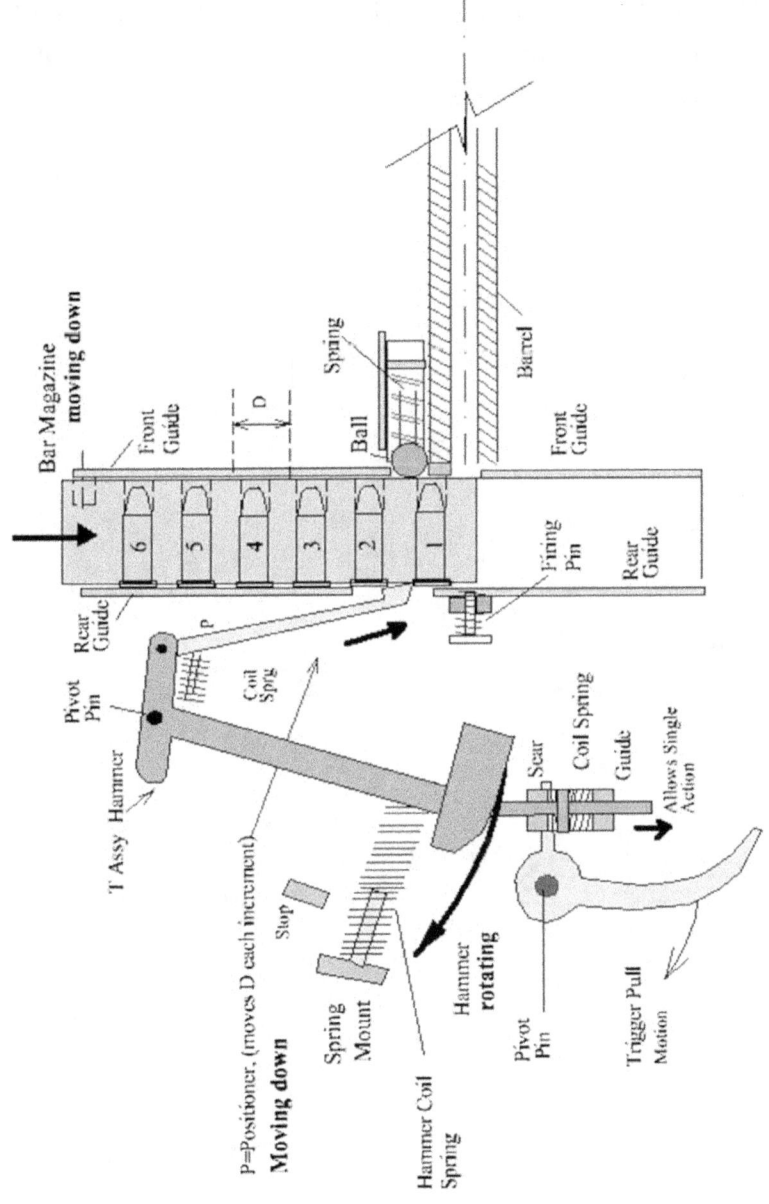

Fig. 11. Hammer Rotating.

- As the rotation continues, the sloped hammer head slides past the Sear latch, as the positioner continues pushing the magazine bar downward with the first round about to align and lock in with the barrel.

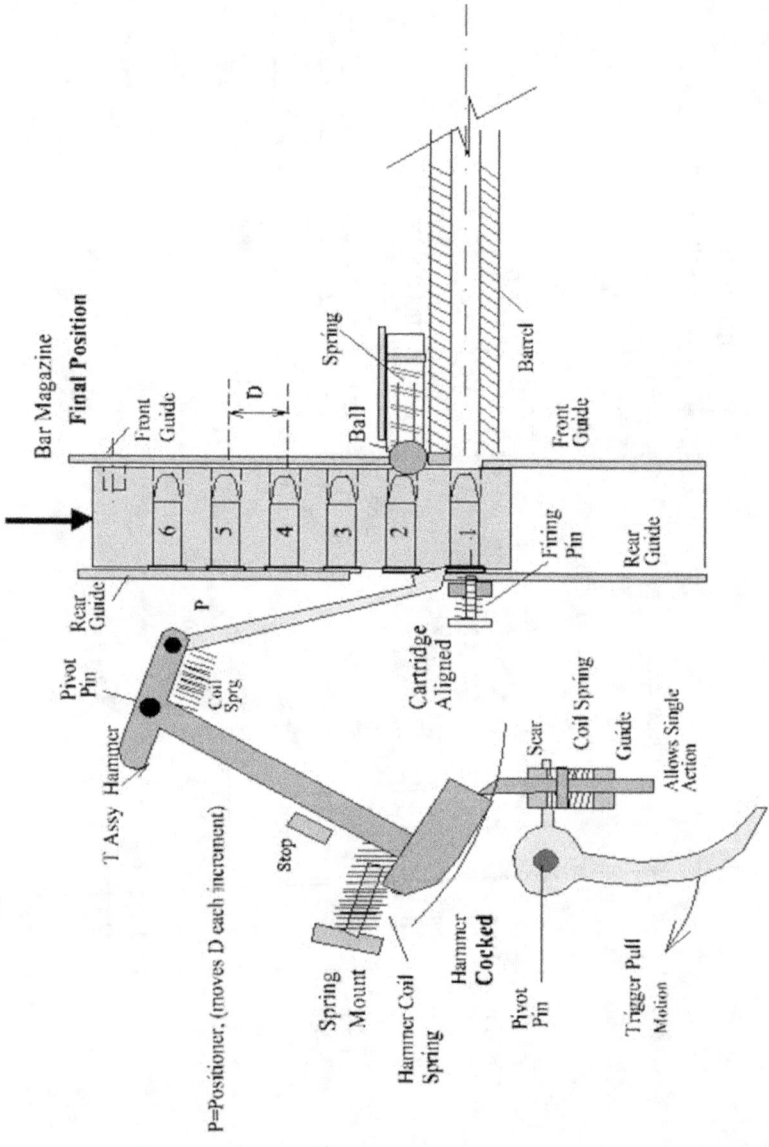

Fig. 12. Pistol aligned to Shoot.

- And then in Figure 12, the Hammer is locked by the Sear as the index ball also locks the Magazine Bar and first round in alignment with the barrel.

This completes the cocking cycle for the simple Single Action mechanism shown.

Look at the simplicity. The only thing the shooter does is pull the Hammer back and the round is in place for firing. There are only two major moving parts that were needed for the operation, the Hammer and the Positioner.

The round is now ready in alignment with the barrel and the Hammer is cocked. Pulling the trigger will fire the gun.

As you can see, for Single action, this is a very simple mechanism.

(This is actually a simplified view of the action, there might be brackets not shown that hold parts in place, like the firing pin, but the drawing gives you the general picture.)

Double Action is more tricky and will not be discussed in this book.

SAFETY

All these tools require care in using, and the use of safety equipment in every operation! Protective Goggles are ALWAYS a good idea, Ear protection, and a dust mask are necessary for noisy or dust producing jobs, grinding metal, etc.

Sometimes leather gloves are needed to avoid a hot work-piece.

Grinding or cutting with the angle grinder is especially dangerous and sometimes throws hot sparks.

And of course welding requires care and proper welding equipment; this is another tool which can have hot debris or a hot work-piece. That means all flammable materials must be kept clear!

Always read the directions and warnings provided with

such tools and practice on scrap before doing any serious work.
BE SAFE!!

Along with being safe, try to keep your work area fairly clean; do cleanup often to avoid misplacing parts or tools, and to prevent any flammable material from being ignited by sparks from grinding or hot parts from welding.

A comment on workmanship too, there are always different ways to accomplish a mechanism and sometimes there is an advantage to one over another. Also there are times when one method has a better setup or is easier to do or is more reliable.

So be careful, pay attention and remember, anything you do is *your responsibility*. If you attempt anything in this book, it is your effort, your work, and your choice.

What is shown here only shows my build and is not a final answer to the design. Other methods also can work, and might be better choices.

Chapter 2

Barrel Preparation

The first thing to do is to obtain the barrel. Used barrels are readily available at firearms and gun shops, and also usually there are a pile of used barrels at a vendor's table at a gun show.

I found a used Ruger 10-22 barrel at a gunsmith's shop in my own town. Prices for common .22 caliber barrels typically vary from $10 to $20.

Sight down the barrel with a light colored paper at the other end to reflect the light. You would expect to see a spiral series of grooves, the rifling, which spin the bullet as it travels down the barrel to stabilize its flight for accuracy and consistency. If the barrel seems straight and has fairly clear rifling, it should be fine.

There are a few small preparations for the barrel:

- The barrel must be cut off to the desired pistol length.

- It must have a "forcing cone" cut into the entry portion of the barrel.

Cutting the Barrel

Wrap the barrel in several layers of paper, or between a couple blocks of wood in a vise, and tighten gently in a bench vise to allow cutting the forward end off at your desired pistol length,

Fig. 13. Barrel Cut to 8 Inches.

say 6" minimum. For my pistol, I elected to cut an 8" barrel.

Clamp the barrel at the breech end just above the portion you wish to use, so as not to pinch the portion you intend to build into your pistol.

Use the angle grinder tool with a cutoff disk installed to cut off the barrel at your intended point.

BE SURE to wear safety goggles, a dust mask and ear protectors for this job! Hold the tool securely, and make

sure no flammable items are anywhere near. As the metal is cut, hot sparks do fly. And the metal will be hot! BE SAFE!

Cut slowly and carefully perpendicular to the barrel axis, so as not to overheat the piece, and allow cooling at intervals. Support the end of the barrel as it starts to sag near the end of cutting. **Be careful, the piece may be hot!**

Once cooled, gently grind the cut end on a bench grinder to touch it up, use small intervals grinding, so as not to overheat the end.

Doing the "Forcing Cone"

The forcing cone is a tapered opening at the entry of the barrel to gently force and guide the fired bullet into the rifled portion of the barrel, truing its travel as it grabs to the rifling. The fired bullet in the process of leaping from the bar magazine is thereby forced to guide into the barrel to align it to the barrel path.

All revolvers have a similar barrel design for this same purpose, this is only copying the process.

There are differing ways to do this, with a manufactured reamer, or with homemade methods. My intent is to do this job with a garage rough approach, and I believe it is adequate for a rudimentary firearm. Reasonably done, it can be fairly good.

Harbor Freight has a set of three coned "Stepless Drills" cutter bits, this is one tool I used. I also used a number drill tilted and rotated in the breech portion of the barrel to lightly cut some inner portion of the bore, enlarging the opening and made a coned wood tool with 300 to 400 grit carbide paper glued to the cone tip to fine grind and minimize any grooves from the number drill.

This produced a coned effect in the rear end of the barrel so a .22 bullet could enter fully and then could be felt to be gradually gripped by the barrel. Not any sort of fine machining job, but I believe a field expedient method will simulate the desired result.

Initially the Harbor Freight cone cutter, and a # drill bit were used.

The cone shaped stepless bit was pushed into the bore and ran in short intervals. Oil is used as a lubricant.

Then a twisted paper towel wet with thinner was inserted between each operation to clean any debris. Testing was done between runs with a .22 cartridge to see if the bullet would enter partially. The stepless bit did some minor work as a beginning.

Next I used a # drill inserted about 3/4" into the bore. I gently applied side pressure to the drill as I rotated it around the lip of the hole, the intent being to maybe skim a tiny bit off the entry hole. <u>Very little of this is needed</u>, and using very little sideways pressure, test with a bullet after a short gentle action and using the twisted paper towel to clean the hole..

If 2/3 of the bullet will enter the bore, that is probably fine. Stop.

Grinding Tool

Make a tapered 1/4" wood dowel using a belt sander, turning a 1" long tapered tip portion, such that it will enter about 1" into the bore. See Figure 14, next page.

Start a small row of glue down the tapered tip and glue one edge of a strip of 300 to 400 grit carbide paper to it. Tape it with some masking tape to hold it until dry.

Once dry, remove the tape and continue the carbide paper on around the cone, cut excess off and apply glue to attach the end forming a cone ended sanding tool; tape to hold while it dries. See figure 15. This homemade tool can be put in a drill and used to carefully smooth the existing coned opening in the bore to help minimize small grooves in the forcing cone. Rotate it round the opening with slight side pressure to apply grinding to the entire interior sides.

As fine grinding is done, use a drop of oil in the bore occasionally and clean in short intervals with the twisted piece of paper towel. Also keep checking with your test bullet. If most or all of the lead bullet will just enter the bore before it starts to fit and stops, that is probably fine. Look down the bore with a white piece of paper held at the end to reflect light and see if the ridges from your coning are not too prominent. If not too bad you are done!

Fig. 14. 1/4" Dowel for Tool.

Figure 16 shows a view of the barrel with the round inserted to show depth. Actually the whole bullet lead just fits inside and contacts the cone. This completes the barrel preparation.

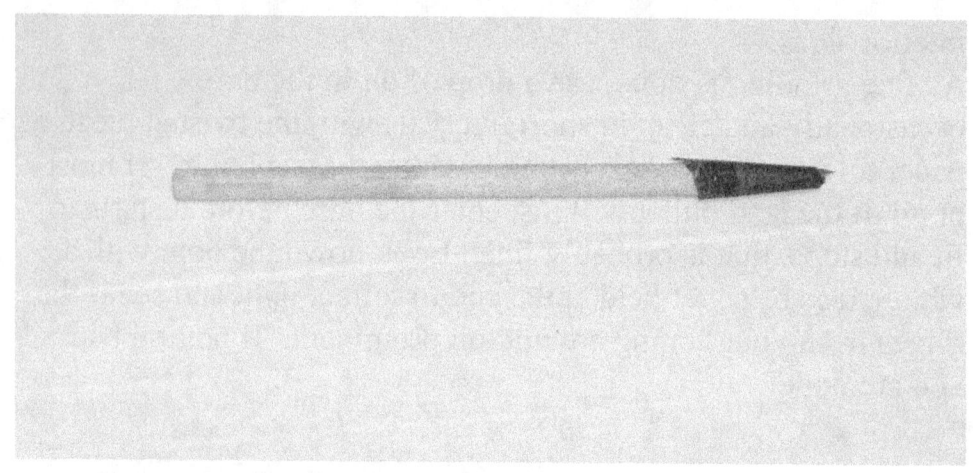

Fig. 15. Tool with #400 Carbide Paper on Tip.

Fig. 16. Bullet inserted in Barrel.

Chapter 3

Receiver and Front Guide

Once the barrel is ready, it is time to cut the receiver panels and the front guide piece.

Look at the the panel dimensions shown in Figure 17 . These pieces are made of sheet steel of about 14 gauge, which you can get from a sheet metal or metal fabrication company; generally they have sheared pieces in their dumpsters, which sell usually by the pound, and quite reasonably too. They are usually a length of flat sheared metal, clean and perfect for our purpose.

You root through the bin and get some 12 and 14 gauge pieces, also a good time to get a piece of material 1/2" in thickness for the Bar Magazine.

The dimensions shown are the dimensions I initially used and once the pistol is planned further, the panels may be modified, as it is often a case that you don't need an area, or can modify an area you don't need.

Cutting the panel is easy and accurate using an Angle Grinder with a cutoff disk. The disks are impregnated with Carbide and are thin, plus Angle grinders are available at reasonable cost from tool suppliers, like Home Depot and Harbor Freight. They typically come in 4" and 4 1/2" sizes.

The grinder is not heavy and is easy to handle, but spins at very high R.P.M. So it is a potentially dangerous device when cutting and requires care and use of safety goggles, Ear Protection and a Dust Mask!

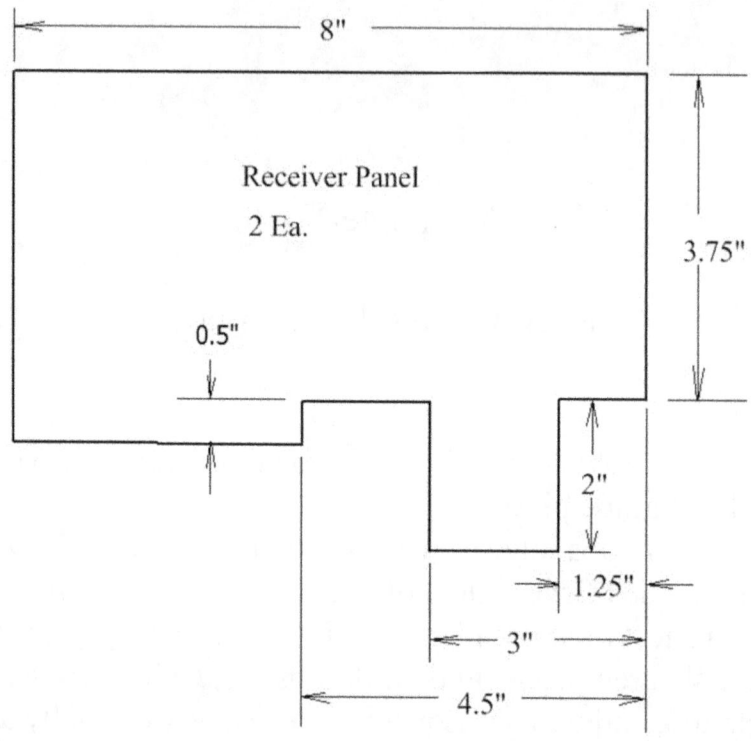

Fig. 17. Receiver Panel, Two Needed.

Do not fail to use caution with this tool. Also in cutting metal, hot sparks are thrown off so avoid any flammable materials anywhere in your work area! Be safe!

Follow instructions on use of the cutting disks, do not apply side stresses, keep the cutting in straight passes only. Practice on a scrap before using on your actual pieces. After practicing a bit you will be confident in using the tool.

Then clamp your metal piece in a bench vise so it is locked solidly in a convenient cutting position.

Check all around your work area for any flammable items and remove if necessary. Place a blocking sheet in the way if it looks like sparks could be thrown in an inaccessible area. Be sure no sparks can go into a dangerous area.

Note all the cuts for the panel are straight cuts. Mark your pieces by using masking tape and a marking pen like a Sharpie. The tape will make the mark easy to follow as you move along a cut line. Work carefully supporting the grinder with both hands and cut each piece accurately as possible. Cut out the two receiver panels.

The Front Guide

Up to now, no careful measurements were required, but now doing the front guide, care must be taken.

The front guide only has two holes, one for the indexing ball, and one for the opening to the barrel. The barrel has to be essentially centered between the two receiver plates.

And in my version all permanent mountings pretty much is to the back (left) receiver panel. The right receiver panel will be removable. I am left handed .. perhaps a right handed person would prefer viewing the right panel with fixed parts... one of those things you must decide.

Either way, in doing the front guide, here is the procedure to follow.

- Using the barrel you have as a reference on width of the front guide. Allow a minimum 1/16" extra width, but cut the height accurately to match the total height of the receiver panels.

- That means the guide height will be 5.75", (5 3/4"...)

- My barrel diameter was 11/16" at the rear cutoff point, so I allowed a bit more than 3/4" in width.

The reason for this is because it is very difficult to do the holes dead center perfect in the front guide when drilling them.

The extra width allows for grinding the sides to achieve perfect centering for the index hole.

The indexing hole must be exactly centered so as to center as well as catch the bar magazine. It must be *centered because the barrel bore is centered and the magazine bar must be in perfect alignment to control that position!* The hole for the barrel bore is drilled centered as possible in the front guide. And the bottom of the barrel should be aligned to the receiver bottom similar to Figure 18 to determine the location.

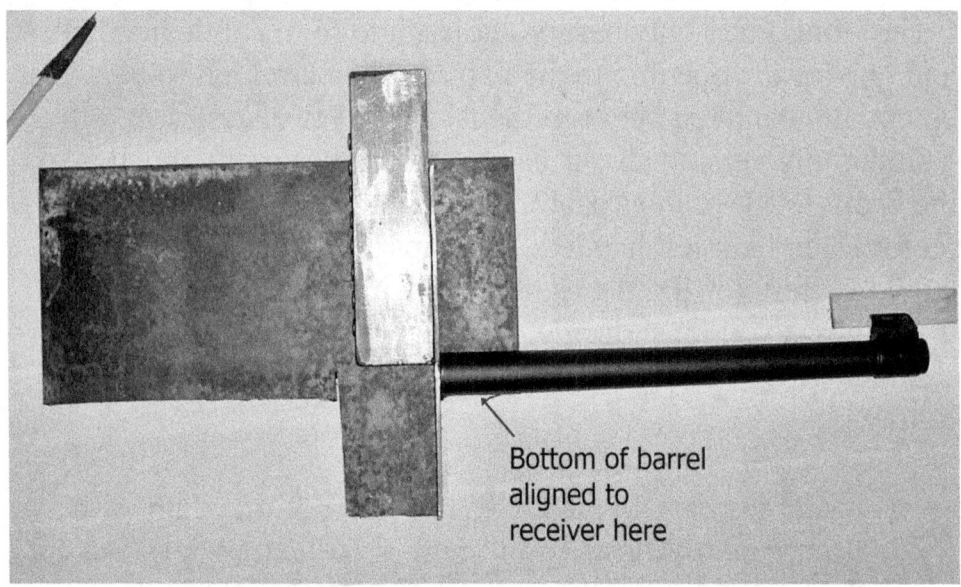

Fig. 18. Align to Set Hole Location.

The hole for the barrel bore is not very critical, and just has to be large enough to allow the bullet to pass unhindered.

Initially we will drill a small hole for both the bore and the indexing ball, then we will use an X-Y table on the drill press to determine the exact repeatable spacing for the bar magazine holes. This insures the accuracy of the bar magazine to match to the bore and index ball settings.

- *A note on drilling: For larger holes use the low speed pulley setting of the drill press. For smaller bits use the higher speed pulley setting. These pulleys allow a large range of speeds. Always use some oil lubrication on the bit as you drill, and do not force the drilling, gentle but steady. And allow cooling in between with the drill press spinning.*

- Set the front guide in place and align the bottom side of the barrel to the bottom front portion of the receiver. Mark the height where the barrel bore hole must be and center drill it, with a smaller drill bit, say 1/8" See Figures 18 and 19 as a guide.

- Looking at YOUR cut-to-length barrel set in place, measure the diameter and visualize a 5/16" indexing ball above it to lock the bar magazine and do a careful estimate as to the distance above the bore to allow the index ball and spring to be housed with a little clearance just above the barrel. ***Probably a minimum of 7/16".***

- Set your front guide on the receiver panel and carefully estimate the distance from the bore to the center of the indexing ball. Using this visualization, you must drill a carefully centered 1/8" hole for the indexing ball.
(For my barrel diameter my spacing from index ball to the bore was 7/16" approximately.)

- So for my indexing ball hole, it will be about 7/16" above the barrel bore hole and centered.

Once drilled, you may find the index hole isn't quite centered. But, because the width of the front guide is a tad wider than needed, you can grind as needed to center and set the width of the guide. The hole for the barrel bore is NOT critical, it later has to be big enough to clear the bore, it can be elongated or larger, it is not critical! The critical centering requirement is for the index ball, not the bore hole. Carefully grind the remaining portion of the front magazine guide to the correct width.

Assuming you too have a similar measurement, what you would now have is a front magazine guide with two 1/8" holes spaced about 7/16" apart. Remember, that distance is based on your barrel. Mine worked out at about 7/16" with the Ruger barrel.

This spacing is the spacing you must have in your bar magazine to match to the front guide, but it must be very accurate.

Here is how you make sure you match your magazine to the spacing:

You carefully mount your front guide in the X-Y Table on your drill press. (Flat face up...) You place the 1/8" drill bit in the chuck.

You will only use one axis as you drill the bar magazine, probably the cross slide is best. Right to left, or left to right, your choice.

- Once you pick your direction, go past the one hole a bit towards the outside edge of the guide, then begin slowly coming back as you check up and down with the 1/8" bit for alignment. Also check that as the slide moves back and forth, the bit stays centered over the work-piece.

- When the bit just slips nicely into the hole write down your cross slide dial reading. (You may need to adjust the other axis to center the bit front to back.)

- Raise the bit and continue in the same direction as you carefully watch the dial on the cross slide. Keep track of the full turns and thousandths. Move slowly and carefully.

- When you reach alignment with the other hole note the difference in readings. This difference is the distance between your two holes in thousandths. Write it down.

- If you wish to repeat, just to be sure, always move in only one direction for accuracy. That way you are always absorbing the slop in the mechanical gearing of the dials in the same direction so it does not upset your measurement.

You will now have the exact distance between the index hole and the bore hole. When you drill your bar magazine, it should match the spacing perfectly.

With the spacing determined between rounds you can now drill the bore hole to 5/16" for extra clearance for the bullet that will exit the magazine.

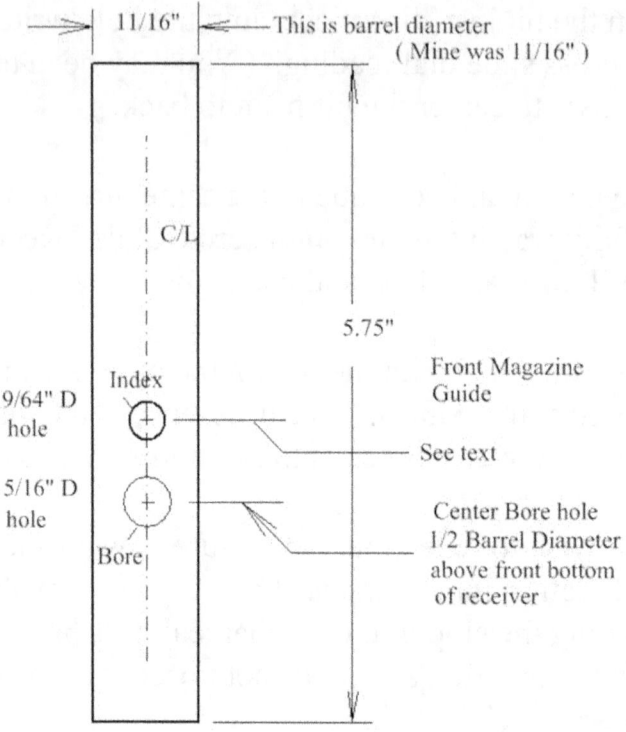

Fig. 19. Front Magazine Guide.

With the front magazine guide in the drill press and X-Y table use the 1/8" bit to align to the barrel bore hole. Then replace the 1/8" bit with the 5/16" bit and re-drill the hole. Set to a slow speed for drilling this larger hole and use oil as a lubricant in drilling.
See Figure 19.

Preparing the Index Hole

The index ball hole must be precise and beveled to allow the 5/16" ball to hang through into the bar magazine area, to grab and hold the magazine bar at each round. Have a steel 5/16" ball bearing for checking as you prepare the hole.

- Place the front magazine guide in the drill press and use the X-Y table to adjust so your 1/8" bit precisely matches the index ball location hole you drilled before.

- With the centering established, exchange your 1/8" bit with a 19/64" bit and carefully drill the index hole larger; use a few drops of oil and gently drill the hole.

- With the hole now drilled, install a countersink bit into the drill press. Very carefully, and only making fine cuts, use the countersink and then check that your ball bearing begins to protrude through the hole a good bit. When it protrudes a bit into the bar magazine area side, remove the guide piece from the drill press.

- Now very carefully use a tapered common reamer to *very gently remove a tiny bit of material. You are working towards perhaps having the ball extend almost 3/32" or so into the side where the bar magazine will reside. Such that the protruding ball portion will grab the hole in the front of the bar magazine.*

- You want to achieve this very carefully, where the ball bearing does not slip through the hole. Work slowly and in small increments to get this right!

Fig. 20. Front Guide in Position.

See Figure 20. This shows the front magazine guide set in place to illustrate the final mounting setup. Once confident of the front guide, I welded it to the front of the receiver. See Figure 21. Later near final assembly, the top of the guide will also be welded.

Fig. 21. Welding Front Guide onto Receiver Panel.

Chapter 4

The Bar Magazine Build

It is now time to do the bar magazine. There are two choices here, to either do it in the garage – or have it made by a machine shop or friend with accurate machines.

When I did my .22 rifle, "Homebuilt Firearms" 2010, I made the bar magazine of a 1/2" steel bar, and drilled the holes for the rounds using a #1 number drill bit on a bench drill press.

- A .22 cartridge develops possibly 24,000 p.s.i. (Pounds per Square Inch.)

- Calculus shows the force in a direction to burst the steel works out to cross sectional area of the cartridge times the 24,000 p.s.i. (0.225 x 0.535 x 24000 = 2863 lbs Force.)

- The calculations of this pressure force on the steel, yielded a force to burst the magazine of under 3000 lbs.

- I used 30,000 p.s.i. For the steel tensile strength. Calculating with the bar metal area resisting the cartridge force provides a strength of 9,000 lb, easily able to resist the pressure from the .22 round. (2 x .135 x 1.125x 30,000 = 9113 lbs Force.)

So, I am sure the steel is adequate. Further, because the shell opens at one end to fire the bullet, I suspect a lower true pressure value. Many times firing the prior rifle I built has shown no problems.

However, doing a good accurate job of drilling the holes centered and true is very difficult on a small bench drill press, even with an X-Y table.

The X-Y Table implies an ability to do an accurate job, but it is not easy to achieve due to the give in the dovetail guides of the table and the lack of stiffness of the bench drill press when drilling steel. Also my sequence of drilling did possibly make it more difficult.

I used the X-Y table micrometer adjust wheels to set the spacing for doing a tiny centering entry hole at each location for the rounds using my spacing I had determined from my barrel dimensions in the last chapter. **A double ended center drill should be used to mark the centers.** These are very rigid and short for stable accuracy.

Then I went directly to the #1 number drill bit to drill the hole, also using the X-Y Table micrometer wheels to try to insure accuracy. This made the initial drilling with the #1 bit rough and awkward. The bit chattered at first. And I could see variations in accuracy as I tried to drill the several holes.

I have thought about this and feel if I had used perhaps two or three gradually larger bits to increase the hole size more gradually (with corresponding less force required at each drilling...) it might have drilled more accurate holes. I may redo a bar magazine to verify, and will have the results later.

So, if you only have a bench drill press, and the thought of this operation makes you pause, you can always have a machine shop make this part of 4140 steel, the same as used for revolver cylinders. Their heavy duty machines will make an accurate and perfect magazine of the very strong and even more adequate steel.

Also you could have the width a good match to fit in the receiver without requiring shimming as the 1/2" thick bar will require.

You have your measurement between the cartridge holes from the X-Y Table and can have it made if you wish.

However, the rest of this chapter shows my build, doing it in the garage with a 1/2" bar of steel... using the bench drill press and the X-Y Table.

Because the diameter of my barrel is large than 1/2", it means my magazine will need shimming to fit smoothly between the sides of the receivers yet with minimal clearance.
You will need to have a fabrication or machine shop cut your metal block if using the 1/2" steel. The sides must be parallel so as the magazine advances down the well it remains stable and does not bind or rock. If the bar is not perfect, you may need it surfaced at a shop.

If your block appears true, it is time to drill the holes for the rounds. Here again is another choice. When drilled on the drill press, there will be some minor ridges in the cartridge holes, and later during firing, the casing will not slip out easily. If you make two magazine bars, you will have another magazine available at least before you would need to remove the casings, by tapping them out.

If you had the bar machined in a metal shop, you could have the holes reamed smooth, which would make the casing slip out easily after they were fired.

When you drill the hole yourself, there are slight ridges in the holes, and these sometimes grab a fired shell casing; they will not simply fall out. Something to keep in mind.

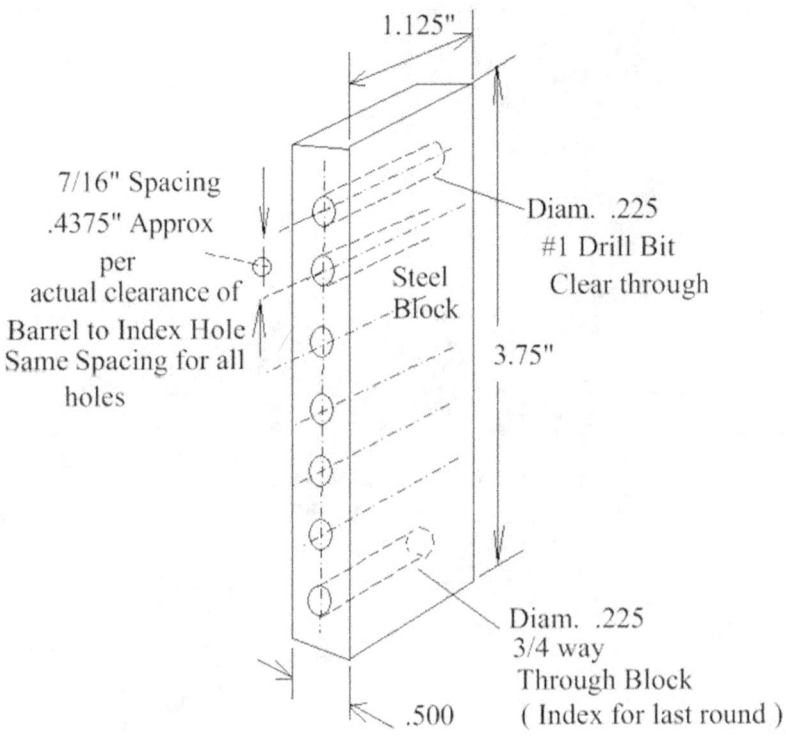

Fig. 22. The Magazine for My Pistol.

The bar magazine will be set up for 6 to 8 rounds, depending on your space between rounds.
The #1 Screw Machine drill bit should be a quality one, available from industrial supply houses. They are short, about 2 3/8" to 2 1/2" long for rigidity.

Order this tool bit online from a reputable machine tool manufacturer.

Fig. 23. Drilling the Magazine.

See Figure 22 for a general idea of dimensions for my magazine. Keep in mind these measurements are for my barrel and fit... the Ruger barrel. Figure 23 shows the drilling for the magazine.

My magazine worked out at 6 rounds; also keep in mind there is one extra partial hole, which is the index hole for the last shot. So you will note seven holes in my final magazine in Figure 23.

Critical Items in Drilling the Bar Magazine

- When you place the magazine in the vise of your X-Y Table, center the center drill in the 1/2" thickness and run the cross slide micrometer dial back and forth to be sure the bit stays centered as you move the bar back and forth, the centering must be accurate as you drill. Set the drill speed at a middle setting. Use the centering drill bit. (Chapter 1. Figure 6.)

- Then, using the round spacing from the discussion in the last chapter on drilling the front magazine guide, do a small start hole into the bar, about 5/16" from one end. Drill about 1/8" in depth, enough to act as centering for the later succeeding bits. Use oil as lubricant. Increment the micrometer to the next locations consecutively and do each hole start, just a slight centering drilled depth. Use oil for each as lubricant. Depending on the table movement you may need to relocate the piece as you drill and recheck the spacing.

- Once all centered, spaced holes are done and are near the other end of the bar, it is time to move to a larger bit. I believe it is a better idea to use slightly larger bits over about three drillings rather than go directly to the #1 bit. I would use increases of about 1/32". Drill the one hole (Last shot indexing hole...) only ½ way through, but drill perhaps 3/4" deep on the remaining holes. Use adequate oil for lubricant. Drill perhaps in 1/8" depth steps with oil lubrication, then back out and repeat, oiling each time. And check with the micrometer feed to verify spacing is still good. As you approach the last larger bit, reset the pulleys in the bench drill press to lower the speed.

- Eventually in the final drilling, with the #1 bit, (Lowest speed on drill press...) drill the hole closest to an end (the indexing hole for the last round...) only 2/3 way through, **so there is no way to somehow load a live cartridge into that spot. See Figure 24.**

Fig. 24. My Magazine Drilled.

- Drill the remaining holes clear through, small amounts at a time with adequate oil.

- The hole that is to take a cartridge at the end of the bar must have at least 1/4" steel remaining at the end beyond the hole.   See Figure 24.

Excess metal at the top or bottom end can be cut off later to cut weight and also allow a partial hammer movement to eject the bar magazine.

One last important item:

- **At the bottom of the magazine in front, grind a slight bevel at the corner to allow the magazine to easily insert against the index ball. Just a slight bit of bevel on the corner will allow the magazine to slide down and engage with the first index hole in the magazine.**

Chapter 5

Finishing the Front
and Rear Magazine Guides

The Front Magazine Guide

When doing the front guide in chapter 3, the index hole was beveled with a countersink and reamer so the index ball would protrude into the magazine area to catch the hole in the front of the magazine as it ratchets through the gun.

Ideally, the ball would index fully into the hole in the magazine as it also rested tight against the front magazine guide hole. But more than likely the index ball sits slightly more into the magazine well than necessary and will have some slop in the front guide hole unless the magazine holes are slightly beveled in front to adjust for a perfect fit. See Figure 25, side view to visualize.

Figure 25 shows that without the bevel, the bar magazine will not be as truly locked to the guide and there will be a poor lock with the barrel alignment.

To insure a good lock, the front holes in the front bar magazine should be touched up carefully with a countersink bit so the magazine **just fits against the front magazine guide with a minimal space as the ball is seated firmly in the front magazine guide index hole**. This is the desired correct situation, and carefully doing each front hole in the magazine will achieve this fit. Once this careful fitting is done, so the fit appears close to the third view in Figure 25, it is time to work on the rear guide.

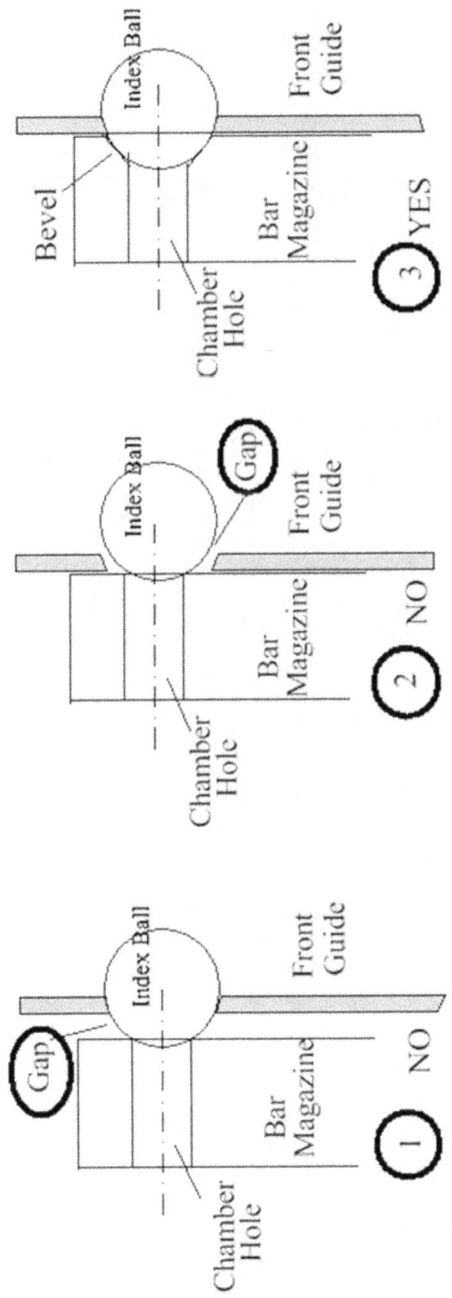

Fig. 25. Side View Index Ball Bevel Setup (3).

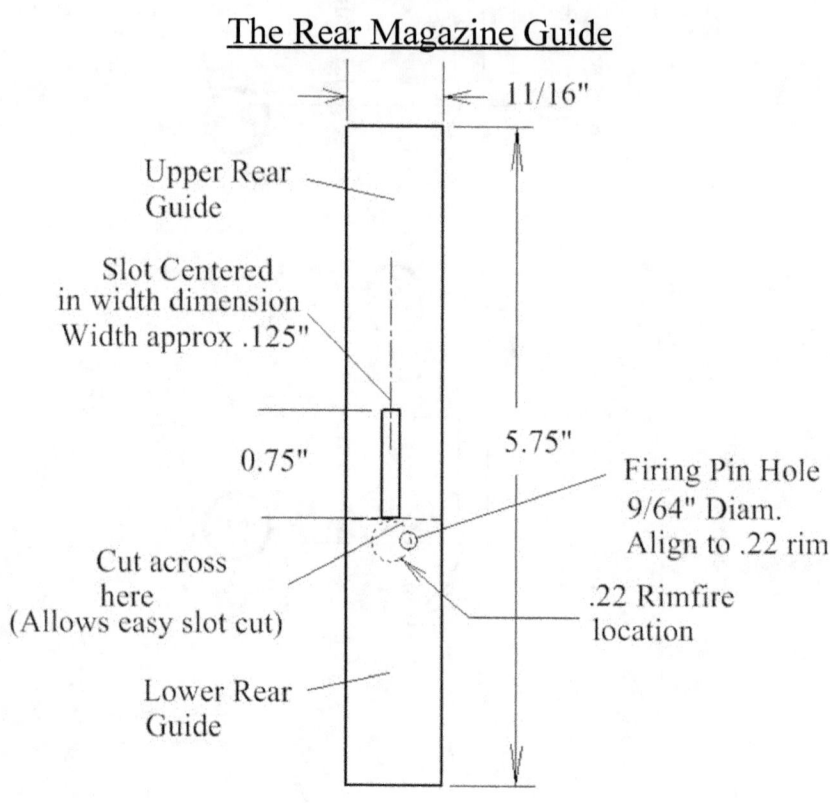

Fig. 26. My Rear Guide Detail.

There are two critical holes in the rear guide.

- The firing pin hole, and,

- The positioner guide slot.

These holes must match to the magazine specific to the barrel and gun. They will vary depending on the barrel diameter and spacing between the barrel and index ball holes.

See Figures 26, 27 and 28, which show the general construction.

With the front magazine guide welded onto the left receiver, as was done in Chapter 3, if you align the bar magazine to the index and bore holes in front, the rear guide can then be located so that the firing pin hole aligns to the rim of the round at the barrel, and the positioner slot is sufficient to allow grabbing the rim of the next round to guide it into position for the succeeding shot.

- Place some fired .22 casings into the holes in the bar magazine. Lay the receiver panel flat on the work table.

- Align the magazine to a pair of index and bore holes in the front guide, and clamp the bar magazine in place. **Shim as needed for centering.** With the rear guide aligned to a fired 22 casing in the magazine directly positioned to the bore, mark the rear guide edge for a firing pin centered on the cartridge rim.

- I set the firing pin hole to align with the rim of the .22 furthest from the receiver panel to fix **one distinct firing position for a round. (Picking a point on the rear guide along the center line of the bar magazine would have two firing points, and if the action malfunctioned, one would be a bad one!!)**

- Also mark the rear guide at the upper edge of the cartridge rim **with respect to the top of the receiver.** See Figure 26. This will be a point to cut the rear guide into an upper and lower piece aligned with where the positioner can push the cartridge into place for firing.

- Splitting the rear panel into an upper and lower part allows careful cutting of the positioner slot with the angle grinder, and also allowed easy access from the lower end of the upper guide piece. Cut across directly in line with the cartridge rim. Use a hacksaw for this cut.

- The vertical slot in the upper guide piece should be about 1 ½ times the cartridge spacing in the bar magazine running up the guide. And it should be centered and wide enough for clearance to the positioner piece which will be 14 gauge steel.

Look at Figure 27 for the appearance.

Also, now is a good time to make a breech block of the 1/2" steel, a small chunk which will be drilled to fit right behind the round aligned with the barrel. This piece must be drilled to match the firing pin hole of the lower rear magazine guide. It can be ground and beveled to fit perfectly behind the firing pin hole, then welded during the final operation to mount the rear guide. See Figure 28 and 29.

This is a very important part of construction and requires care and accurate drilling.

The bar magazine or the area it slides into can be shimmed as needed to do a perfect positioning for the alignment.

Take the time to do this right. This is an important part of fitting the magazine.

I do believe wood or metal shimming can do this job.

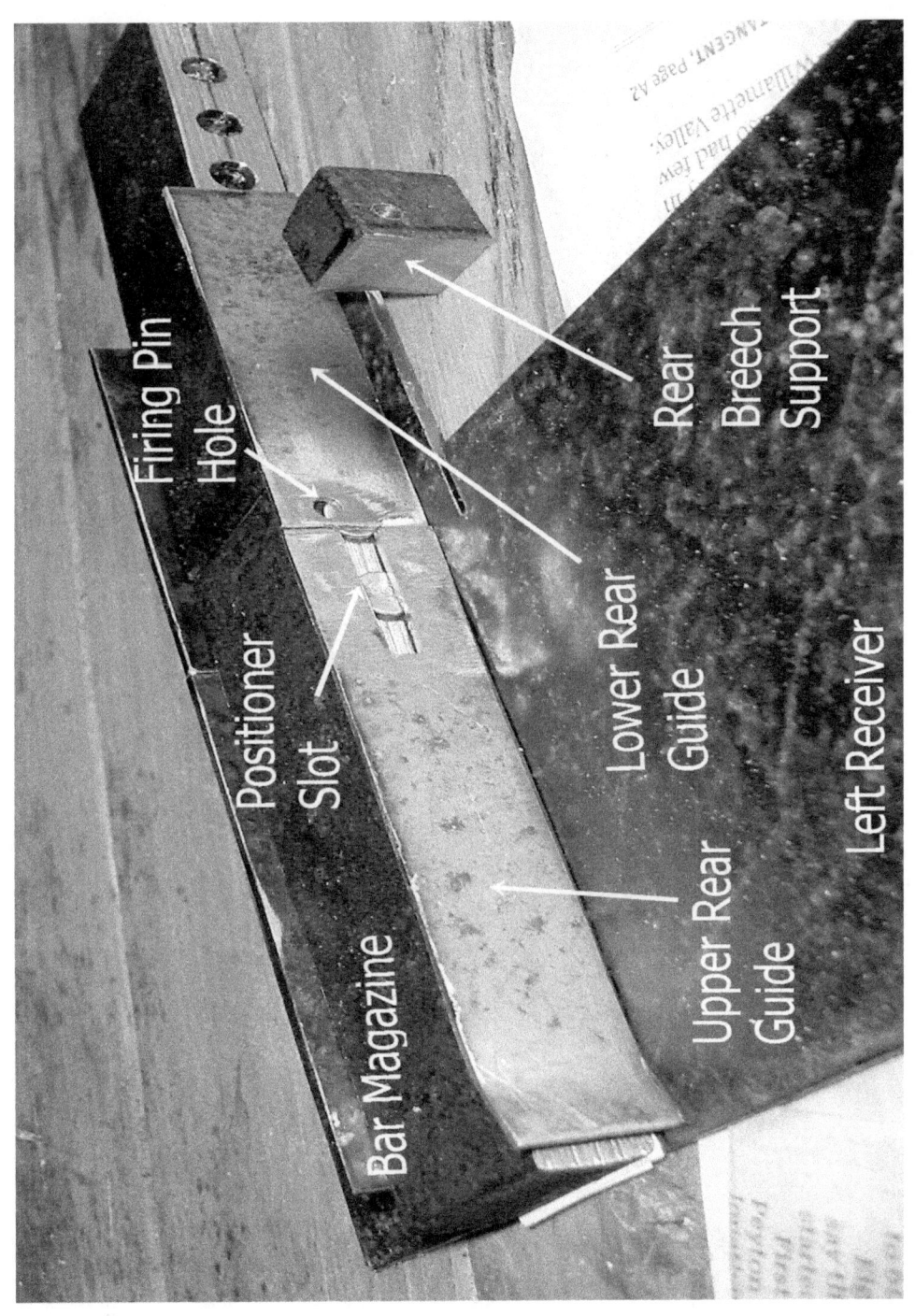

Fig. 27. Critical Rear Guide Features.

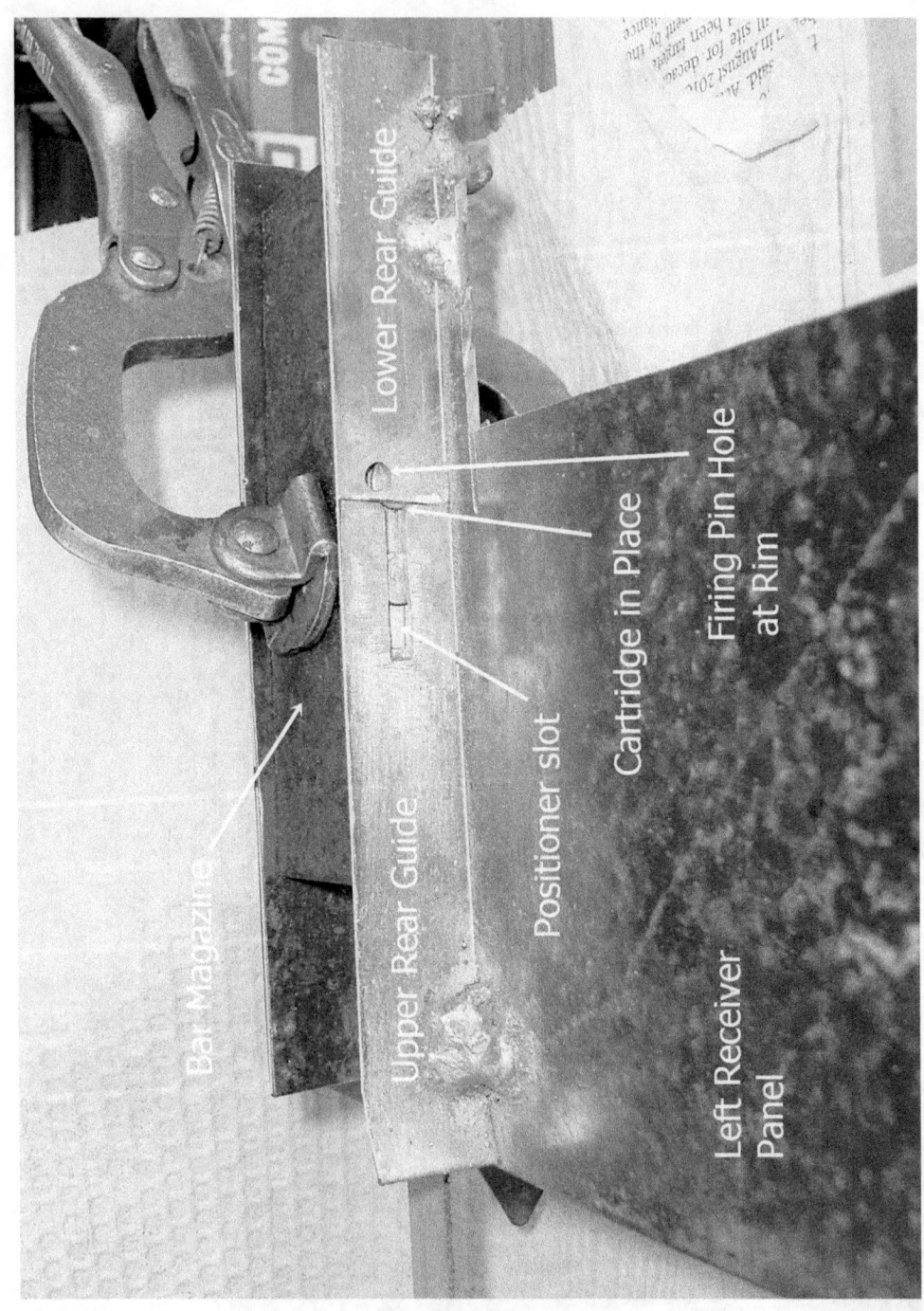

Fig. 28. Rear Guide Welded.

Fig. 29. Firing Pin Block Welded.

Index Ball Bracket

Bend a small "L" bracket to contain the index ball and spring at the front magazine guide. Weld it to the receiver in front of the front magazine guide as required to apply a spring force to the index ball. Make sure it does not protrude above the height of the magazine guide from the receiver surface.

See Figure 30. The spring I used here was from Hillman hardware parts bins or Amazon, Reed Industrial or similar types, many similar: Index Ball Spring: 1 1/4" L x 3/8" O.D. .040 Music Wire.

The bracket is approx 1" from the front guide to provide compression against the index ball.

Fig. 30. Bracket for Index Ball.

Chapter 6

Making the Internal Operating Parts, and Testing

This chapter discusses the determination of operating parts and positions of the hammer, positioner and other parts to provide the operating setup for the pistol.

- The hammer must have a spring driving it sufficient to fire the cartridge.

- The firing pin must be confined in place and provided with a light return spring.

- The cocking of the hammer must also provide the positioner with the exact correct movement to position consecutive rounds for firing.

- The last partial movement of the hammer must eject the magazine.

There is an order to doing these setups, first the hammer-spring swing arc must be verified capable to fire the cartridge.
(Thus, both the firing pin arrangement and hammer unit must be built first.)
The firing pin hole has already been done. Once that swing

arc is known, the upper clevis arm radius operating the positioner must be determined to insure the correct coinciding positioner movement.

The Hammer

Figures 31 and 32 give the general picture for the hammer. As the hammer rotates around its pivot point, the clevis at the top right also rotates an identical angular amount. The distance the positioner moves must be equal to the spacing of the magazine holes, plus a small gap to insure the position rests just above the next cartridge rim after firing of the prior cartridge. The final cartridge location as the hammer just reaches its full cock position should align the bar magazine with the front guide index ball, which should lock the round in the correct spot. Visualize this as you study the diagrams.

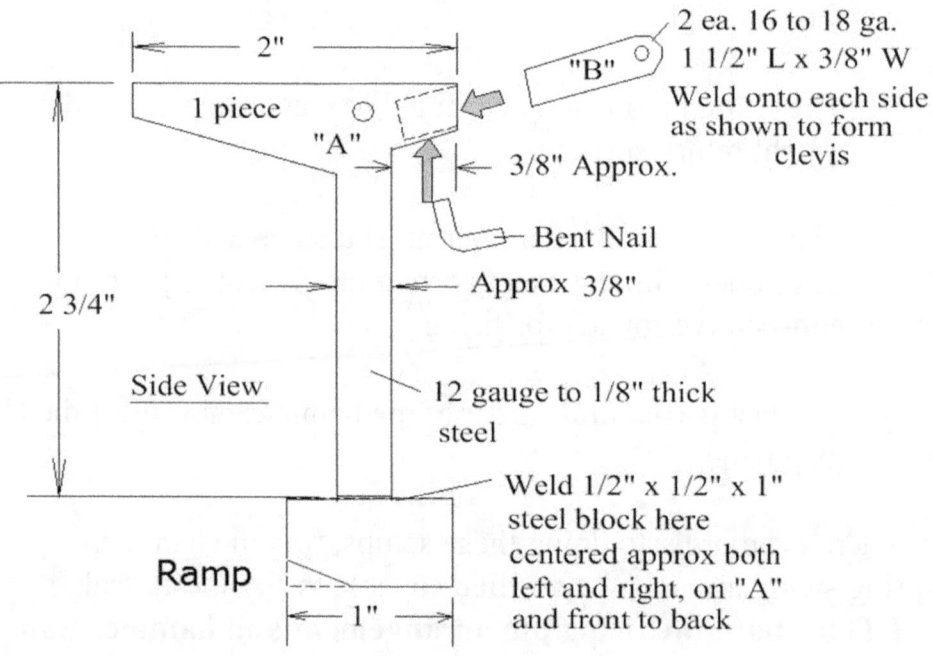

Fig. 31. General Hammer Assembly.

The hammer assembly is made of a "T" shaped piece of approximately 12 gauge to 1/8" thick steel. At the top right portion of the T will be two thinner steel pieces, approx 14 gauge, tack welded to the "T" to form a clevis.

At the bottom is a chunk of steel, the hammer block, about 3/8" to 1/2" thick, welded to the "T" assembly.

Round the right corners of the clevis, so corners are not sharp.

Fig. 32. Photo of Actual Roughed Out Hammer Unit.

Figure 32 shows the actual piece. The right clevis arm of the "T' angles slightly upward. It is not yet finished, but you get the idea. Actually, the left top portion of the "T" hammer can be left off; it was originally for a possible double action version. The front clevis portion is to operate the positioner advancing the bar magazine.

As mentioned before, the first thing to determine is the arc movement required to fire a .22 cartridge in the bar magazine.

So, it is time to do the firing pin next.

The Firing Pin Build

The firing pin is made from a 1/8" High Speed Steel drill bit from the local hardware store.

- It should be put in a bench vise and cut with the Angle Grinder so that the shank and a bit of the flutes will be long enough for a firing pin. I did mine 1 7/16" long.

- A .22 firing pin normally impinges on the rim portion of the round, and is a very narrow surface at the point it hits the rim. So the bit of fluted drill end opposite the shank should be ground to a very narrow bladed tip like a fine point blade screwdriver. See Figure 33.

- The firing pin will have a return spring and something to retain it from falling out from its guide hole in the rear magazine guide. This requires a rear guide in addition to the rear magazine breech block done in the last chapter.

- This rear guide can be a simple angle bracket. However, the firing pin must have a metal tab piece welded on to provide a spring flange and also a means to hold alignment and keep the pin from falling out.

- Keep in mind, the firing pin is a small piece of 1/8" drill bit. So the weld here is a very tiny, tiny speck of fine welding. It has to be welded, the impact of the hammer hitting the firing pin would jar epoxy or any other glue to failure.

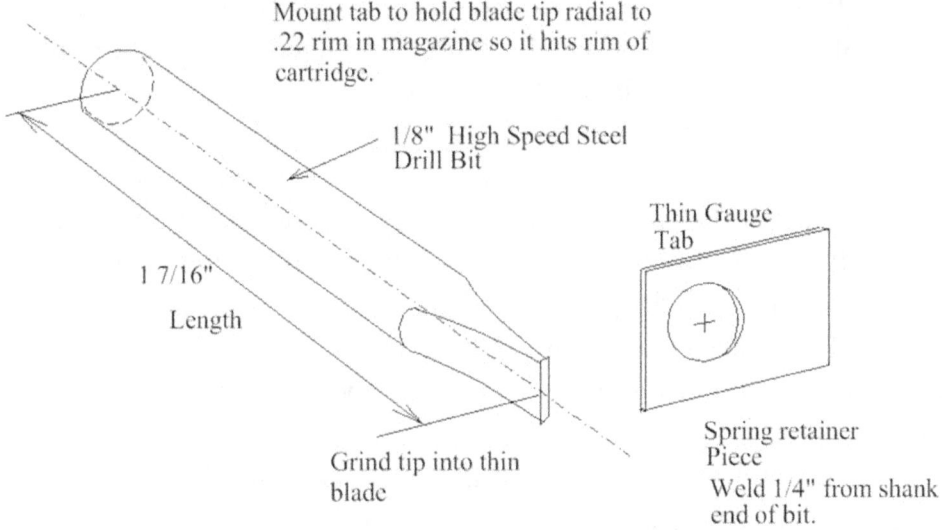

Fig. 33. Construction of Firing Pin.

- So, look at Figures 33 and 34 to see the final setup for the firing pin.

The firing pin retainer is a small metal rectangular piece to:

- Keep the firing pin from rotating as it has to hit with the blade end radial to the cartridge center. It has to be welded with just a tiny spot to hold it, and it is meant to keep the bit from rotating out of alignment. If it tries to rotate, the longer side of the retainer should hit the receiver sides to prevent significant rotation.

- Also this piece retains the firing pin spring. Your receiver spacing and the screw mounting the rear guide must be kept in mind to achieve this. Some grinding to clear the firing pin rear bracket mounting may be required. The rear stop bracket can be a simple "L" bracket. See Figure 33, which shows the parts in my version.

Push the firing pin into the block on the rear of the magazine guide. Sit the rear firing pin bracket onto the receiver panel with the mounting hole to the right. Slide it onto the firing pin. Mark and drill a 9/32" hole back from the firing pin block to allow about 1/8" of the firing pin tip to enter the bar magazine area while still guided in the rear L bracket and about even with the left bracket surface. This will insure a good hit with the hammer to fire a cartridge.

Note the tiny spot of weld on the pin, also note the small spring. This spring is from the Ace assortment.

A Note on Springs

Hardware stores which have bins of small screws, bolts, springs, and such will have various assorted springs in a variety of lengths and strengths. The stiffness of a spring is determined by its thickness and material, and most of these springs are either Music Wire, or Stainless Steel.

Besides local hardware stores, Amazon has an Industrial and Scientific section of spring products.

The spring for the hammer, and the index ball are heavier stiff springs, typically about .040 diameter wire, about 2 to 2 1/2"

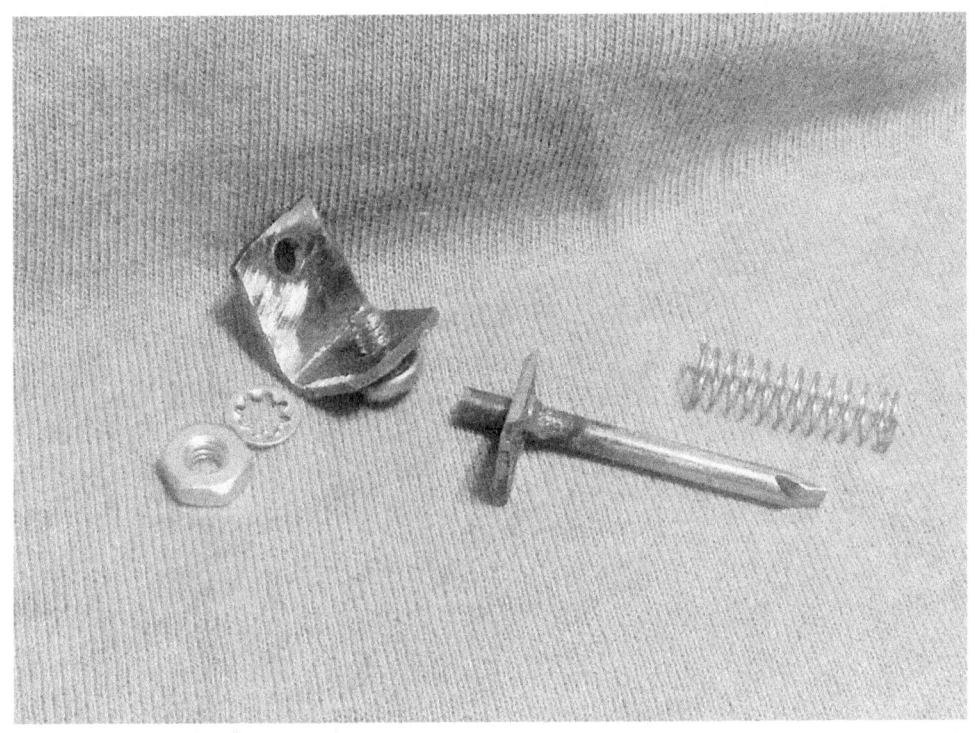

Fig. 34. Parts for Firing Pin.

long for the hammer-spring and 1 1/4" for the index ball spring.

Ace hardware and others have some lighter spring assortments as well. One small Ace spring assortment will have good springs for the positioner, and the firing pin return; these are smaller, less stiff springs. (I will try to list the ones I used once testing shows the best ones for my build.)

Parts Mounting and Details

See Figures 35 and 36 for an actual mounting view of the firing pin, bracket and spring.

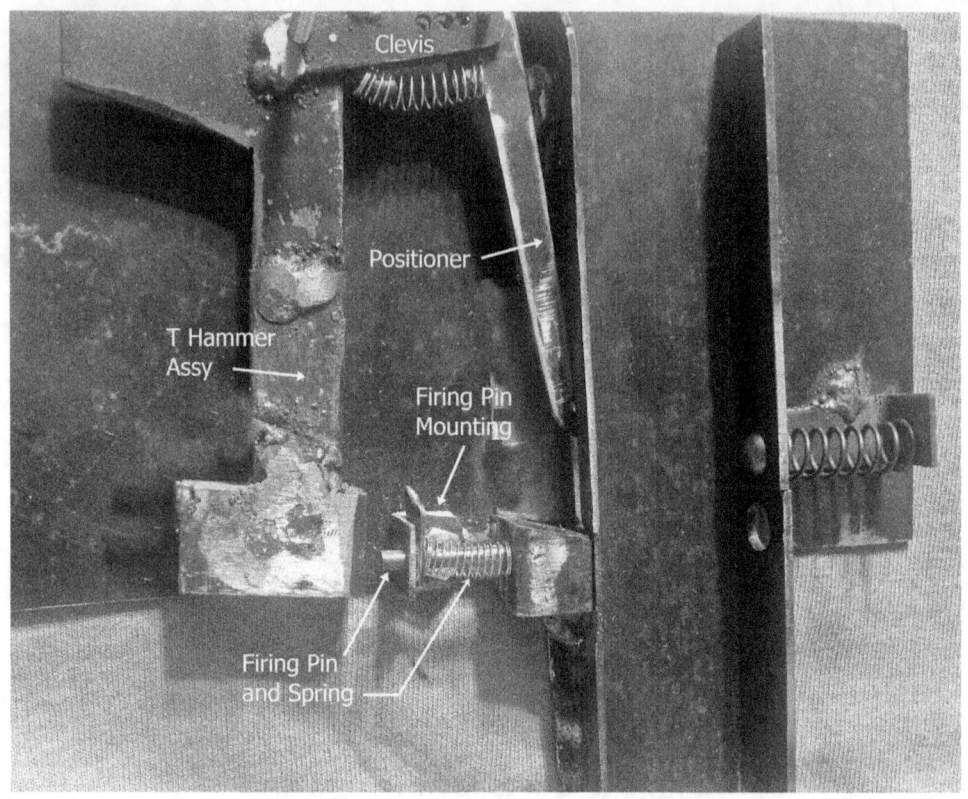

Fig. 35. Showing Mounting of Firing Pin.

The firing pin rear bracket must mount with the fastener beneath or above the spring and pin so it does not interfere with the hammer strike. This may require a bit of grinding on the firing pin spring retainer in the area where it could hit the nut of the "L" bracket fastener... it must clear, but the farther edge of the rectangle must still be such to keep the pin from rotating once the sides of the receiver are in place. And the firing pin parts must be removable in case of any problem.

It is time to determine the pivot point for the hammer on the receiver panel. This is fairly easy to do. Set the "T" hammer unit on the receiver panel and also place the firing pin in the

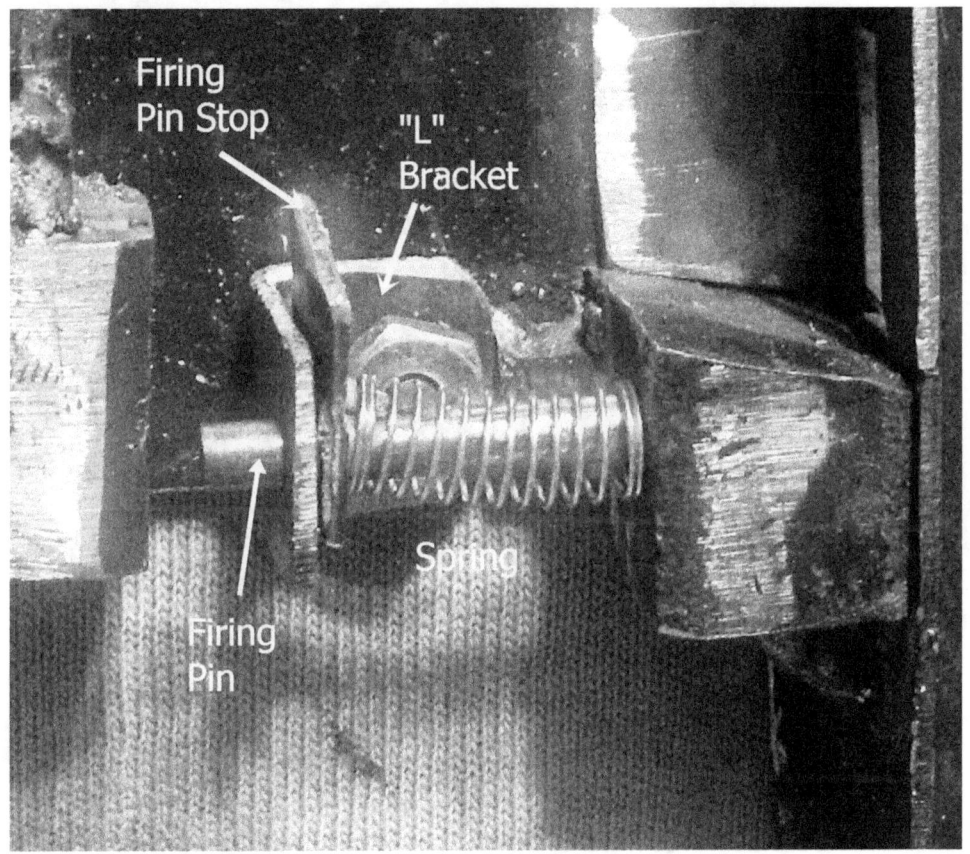

Fig. 36. Firing Pin Mounting Detail.

block portion of the rear magazine guide.

Align the hammer block near the bottom of the receiver panel in line with the firing pin such that it would strike it nicely yet not extend below the receiver panel. Fairly even is fine.

At the top, the hammer unit should sit below the top of the receiver panel with the clevis portion at perhaps 1/2" clearance from the rear magazine guide. Imagine it rotating around the pivot point, and be sure there are no welds or impediment to it rotating freely. You will probably come up with a pivot point

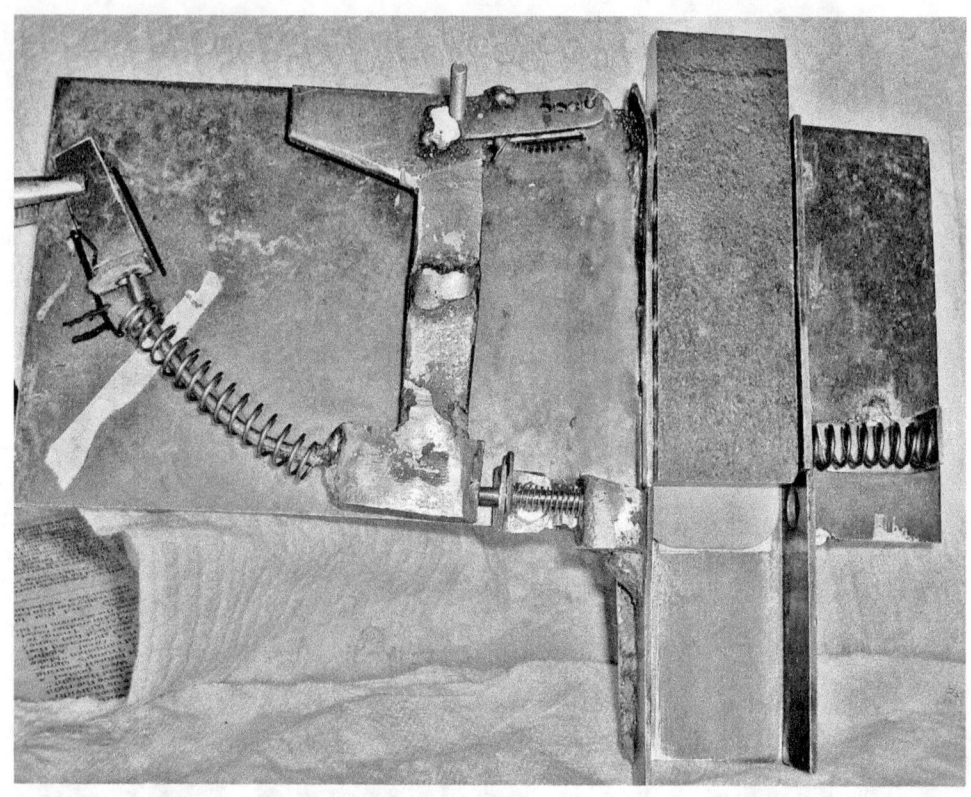

Fig. 37. View of Hammer Spring Test Setup.

about 1 1/2" from the rear magazine and 3/8" down.

Once you are certain, drill a 9/64" hole in the receiver panel at this point. Run a 1" 6-32 screw up through the hole and lock with a nut and lock washer. This will be your pivot point for the T hammer assembly.

Next, look at Figures 37 and 38. In installing a hammer spring, it will be found some sort of guide is required to keep the spring retained, yet not interfere substantially with its operation.

I decided on a guide attached to the rear of the hammer block, and fitting through a large hole in a spring stop bracket. I clamped a #8 nail to the rear of the hammer block, and welded the nail head to the hammer block.

The spring stop is simply a strip of steel bent to an " L", 12-14 gauge is fine, bent into the L shape. See Figure 38 for general details of these parts.

- The hole in the spring stop should be large compared to the #8 nail so as to allow full rotation in a cocking motion of the hammer without dragging against the nail guide. Perhaps 1/4" diameter. At this point in the project, a builder should be able to set the size of this piece.

- The center height of the hole should be made to center between the two sides of the receiver.

- And the nail welded to the rear of the hammer block should be centered in the block.

- To determine the curve of the nail guide, lay a piece of paper beneath the T hammer unit. As you rotate the hammer, align the pencil to the nail head where is is welded to the hammer to trace the curve as you swing the hammer.

- Using the paper as a pattern, clamp the hammer unit in a bench vise and carefully bend the nail to match the curve.

Once the curve looks fairly close, replace the hammer over its pivot, and temporarily sit the spring stop on the receiver and test for your spring guide movement and clearance to the spring stop as you move the hammer. Adjust the stop position and guide curve to achieve a smooth rotation with no rub on the stop hole. Also keep in mind you must test eventually that the arc will be sufficient to fire a round once a spring is installed. Probably 40 degrees or so. See the light colored masking tape strip in Figure

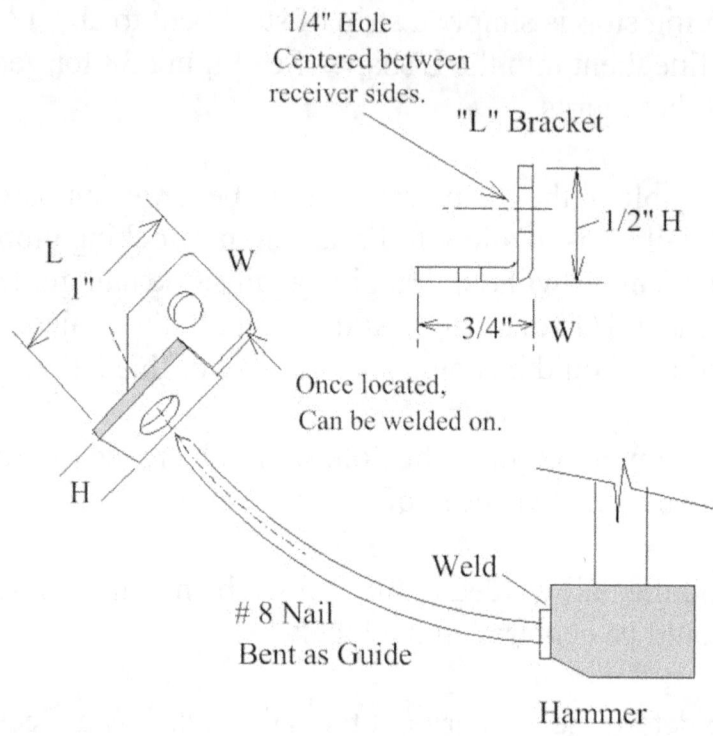

Fig. 38. Hammer-spring Guide and Stop

37 for an idea as to the required arc. (Left of the hammer...) Later on, you will have to determine the clevis radius arc for full movement of the Positioner, as this is the requirement for positioning the magazine to match cocking and so on.

Comments on Construction

On several occasions, in building various parts and mechanisms it becomes obvious there are many ways to achieve a particular function. A catch or latching device or method of holding a piece in place has several options usually, and in some cases one method does take less parts, less construction, fits or works better than another.

In some cases there may be space problems, or other reasons to use one method over another. But if you think about it you can usually come up with a variety of ways to solve the problem. So keep this in mind as a rule during any build. There are always more than one way to do a job!

For example, look at Figure 39. The hammer is rotated to cocked position.

The release shown in Figure 39 is not a bad one, either the hammer or the release rod can have a ramp, and as the hammer rotates, the ramp pushes the rod down against the spring. The rod guides are mounted to the receiver at the appropriate position. As the hammer rotates past the rod catch, the rod pops up due to its spring, and retains the hammer at the cocked position. A downward pull on the rod then releases the hammer.

This last method is however not best for spacing in my build version. But fine for testing.

Further Modifications to the Hammer

At this point look at these other modifications to the hammer. I cut a ramp on the rear of the firing pin hammer, and also added two bosses on the upper pivot point to stabilize the hammer side to side swing more.

Also welded a #8 nail tang below the clevis facing right. This is to retain the spring later that pushes the Positioner towards the rear magazine guide.

Drill through the added metal after you weld on each boss. These are merely two small blocks of steel tacked onto the side of the pivot point, basically thickening the pivot point to fit between both sides of the receiver. See Figure 40 for a detailed view.

Because of the increased sideways width of the pivot portion, the hammer will be more stable in side to side wobble, without the tendency toward dragging against either side. Grind the bosses on either side to center the hammer between the sides of the receiver, and make sure they clear slightly and don't rub.

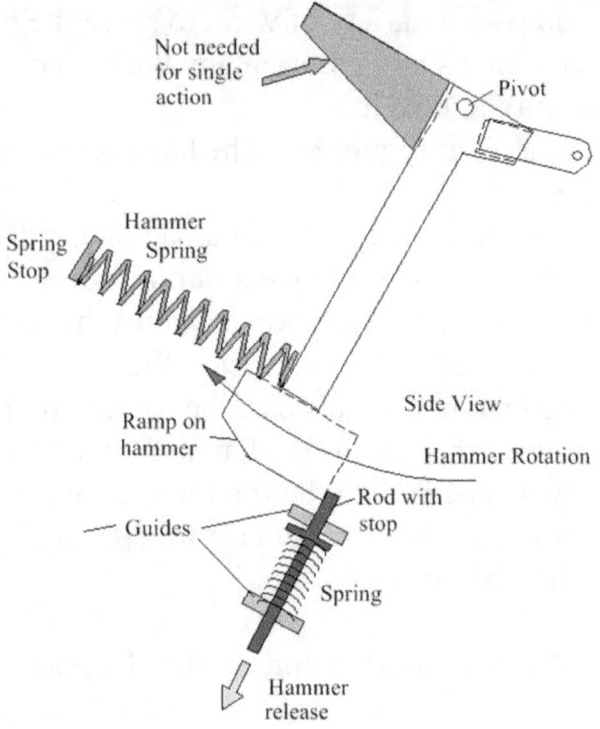

Fig. 39. Sear Catch.

Later on, the clevis holes to mount the positioner can be accurately drilled and set, for now this is not necessary.

Look carefully at the "T" Hammer photo in Figure 40. Note the small nail nib welded beneath the left end of the clevis. This will hold one end of the spring which eventually will push against the positioner.

Note the cocking handle for the hammer. I changed my mind as to showing a double action version of the pistol. The upper left hand portion of the "T" is not needed for a single action mechanism. The simple single action setup is

Fig. 40. Actual Hammer Assembly.

enough for this pistol version; it works very well and is very simple. Weld a 3/4" to 1" length of 1/4" to 5/16" steel rod to the right side of the hammer head as the hammer cocking handle.

Use the angle grinder and cutting disk to cut the gentle ramp at the rear of the hammer weight as shown. These modifications can be verified in Figure 40, which shows each change clearly.

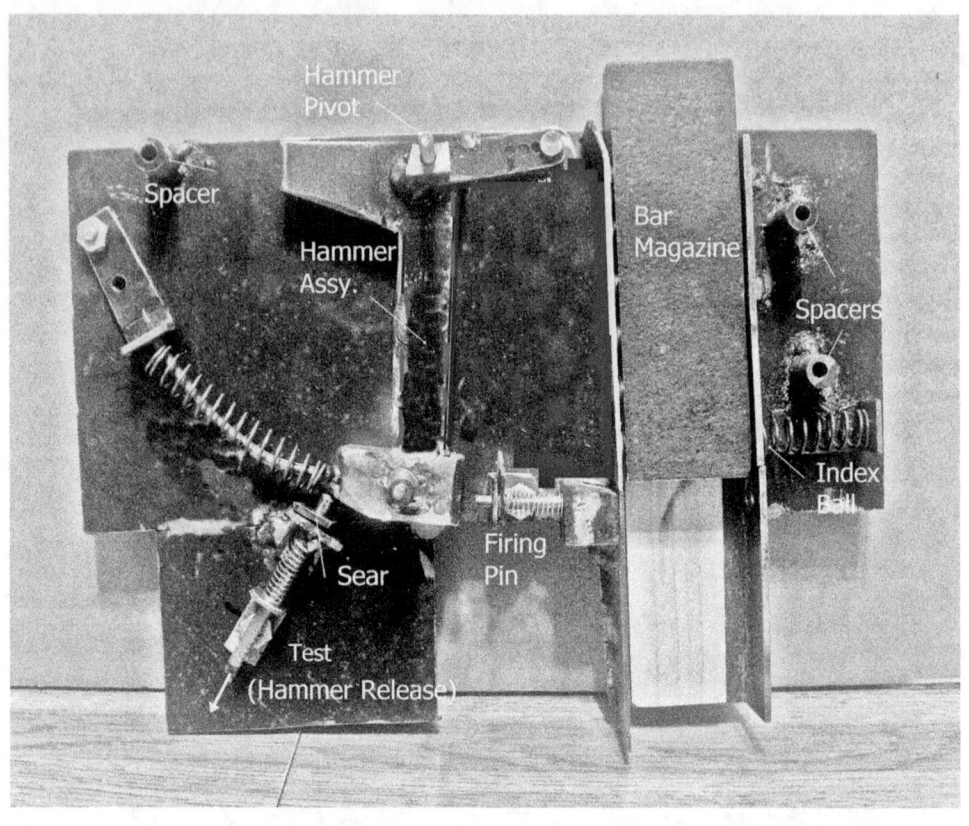

Fig. 41. Hammer and Firing Pin Mounted.

Checking the Mechanisms

We now have a build with the firing pin, and hammer and spring ready to test for proper firing of cartridges in the bar magazine. (I did a temporary test hammer release also.)
With the index ball installed, the magazine will lock in a shooting position for any round. The firing pin should line up for a rim hit on the round at a point farthest from the back receiver panel looking at Figure 41. That will be the .22 rim edge closest to you as you look down at the left receiver like in the picture.

Some mounting spacer bosses are shown in Figure 41, but ignore them for now, they are not yet fully described and are not needed for the test.

The springs I used are listed below:

- Index Ball Spring: 1 1/4" L x 3/8" O.D. .040 Music Wire. Bracket is approx 1" compression.

- Hammer Spring: 2" L x 3/8" O.D. .040 to .042 Music Wire.

- The small spring on the firing pin is a silver colored spring of 1" L x 3/16" O.D. From the Ace Hardware #1 Spring Assortment. A very small minimal force spring.

Checking Alignment

Place the index ball into the index hole and place the spring in place to lock the magazine in position for the first shot.
With side shims as needed, slide the empty magazine bar in until the magazine locks with the index ball in the second hole. The first hole should then align with the barrel hole.

Look into the barrel hole and see if the magazine hole looks fairly centered in place. Check that the magazine seems tightly locked. Use a .22 cleaning rod to check it goes perfectly into the barrel hole and magazine bar.

Continue moving the magazine down one hole at a time as you check the alignment and fit.

If all seems fine for each chamber in the magazine bar, it is time for a test of the hammer and firing pin.

Testing the Mechanism

1. Take six .22 long rifle cartridges, standard, like Remington, Federal, or Peters...
2. With a pair of pliers, twist off the lead bullets and dump the powder and bullets into the waste basket.
3. You should now have six "blanks" which will make a small "pop" and puff of smoke when fired but not actually shoot anything.
4. Load them into the rear of the bar magazine, and head for your shop. **Be sure you have the partial drilled hole at the top front, with the cartridges loaded into the rear of the magazine.**
5. Shim the magazine if needed to center it and slide it down into the magazine guides so the first hole indexes with the index ball.
6. Examine your mechanism... there should be the freely moving firing pin, not showing in the magazine area, but able to protrude inside when pushed from the back.
7. The hammer should move freely on the pivot, and the hammer spring guide should slide through the hammer spring stop bracket without any resistance if the hammer spring is not installed.
8. Temporarily remove the spring stop bracket.
9. Place the 2" long hammer spring on the nail guide and re-install and tighten the spring stop bracket.
10. Now as you force the hammer back around its pivot, you will see it has significant force available.
11. Clamp the receiver upright carefully into a vise so you can operate the hammer for testing.
12. Push the bar magazine down until it indexes and aligns the first cartridge with the barrel.
13. With the receiver vertically mounted, carefully pull the

hammer back against the 2" spring, about a 40 degree rotation, and release it.
14. It should strike the firing pin, and hopefully you will get a "Pop!" indicating it fired the blank!
15. Advance to each succeeding hole and check that each round does indeed fire.

If everything functions normally you are done with the test. Mark the hammer swing position with a piece of tape. This will show the full rotating movement required to fire the rounds.

Chapter 7

My Usable Hammer Release Mechanism

As discussed in the last chapter, there are always several methods to do an operation, and the one used is dependent on many factors, space, availability of material to do the job, etc.

In my pistol, there would not be much space below the hammer for a hammer release like that test version shown in the Figure 41 photo, but there is enough for a different release mechanism.

See Figure 42, next page . A very simple mechanism.

A pull on the trigger will be clockwise, as Figure 42 shows and any rotation of any part of the trigger piece will rotate in that same manner around the pivot. That rotation has to activate the Sear to release the hammer to strike the firing pin.

Figure 42 shows a simplified visualization of the release used in my version. It is just another way to use the trigger rotation to pull the Sear downward to release the hammer. The spring strap holding the Sear in position to lock the hammer is simply a piece of pallet banding, good spring steel which can be cantilevered out from the handle grip of the pistol.

This method of release mechanism fits in my version of the pistol.

As the trigger rotates, it pulls the Sear in a counterclockwise rotation around the right pivot pin and releases the hammer.

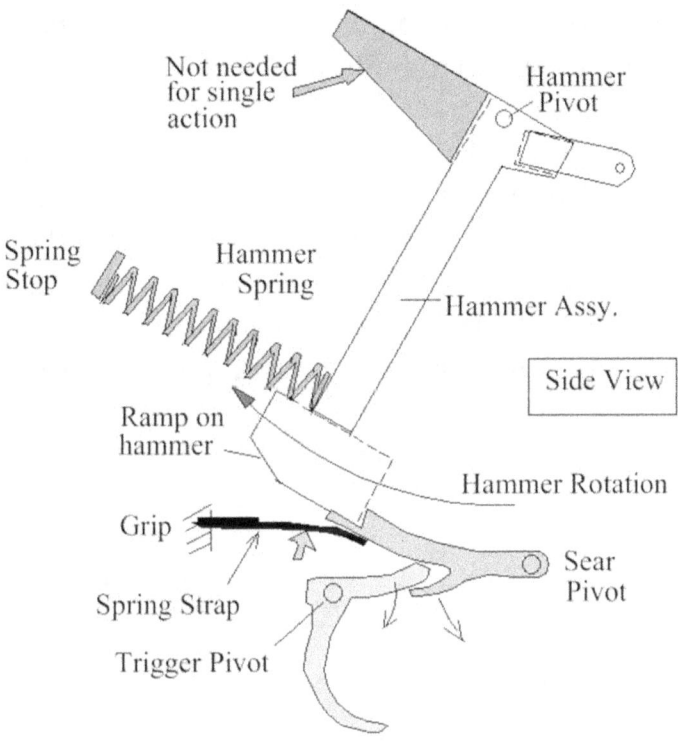

Fig. 42. The workable Hammer Release.

The trigger and the Sear are made of 1/4" thick steel . The drawing is a close representation to actual operation. The left end of the Sear must have an extended step so when it catches the hammer, it is limited how far it rises to block the hammer.

Also the stepped extension gives a surface for the steel spring strap to push upward on the left lock portion of the Sear.

The far right portion of the Sear has a pivot point. This will be secured by a pin in the receiver.

Partway to the left of this pivot is a point where the trigger fits into a lower "V" slot in the Sear to pull downward on the piece, rotating it CCW to release the hammer.

This link point has more trigger leverage the further it is from the Sear pivot. The upper part of the sear is ground so that it just clears the hammer as the hammer is cocked. The Sear does not have to be very thick top to bottom.

The alignment of the rotating parts is such that the point where the trigger rides in the Sear "V" has to allow for two opposing arcs during operation. This means the trigger extension piece will make some back and forth movement in the slot during rotation. But the slot allows the trigger extension to slide forward or back as it operates the Sear movement.

See Figure 43, which shows a general parts description.

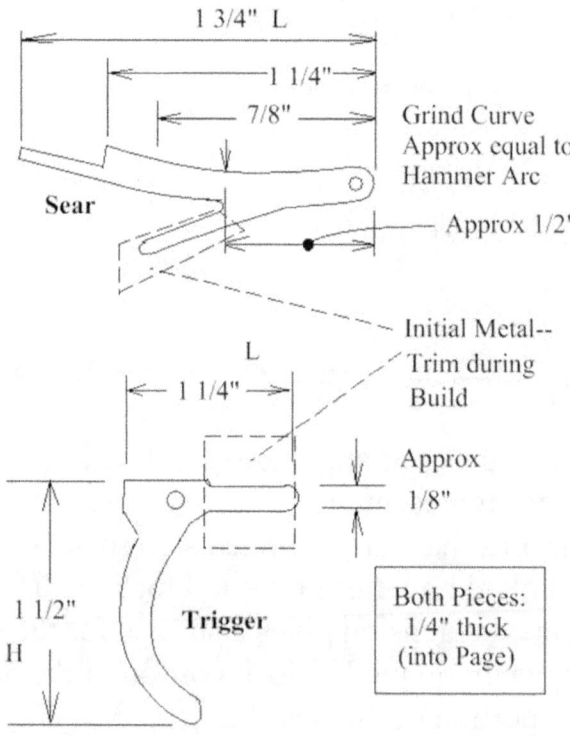

Fig. 43. Parts, Trigger and Sear.

No curvature notes for the trigger are shown in Figure 43 as the builder should have an idea as to his desired fit for this. As long as it is about 1 1/2" height, it should allow a nice space to do the curve.

The pivot hole at the upper end of the trigger and at the right end of the Sear is drilled with a 7/64" bit in the center of those areas as shown.

Two #6 Box Nails will be used as the pins in these areas. This size pivot pin is quite sturdy .

The Trigger

****Use safety goggles, ear protectors and mask ****
Be Safe!!

Cut out the trigger piece rough from a piece of 1/4" steel, using the Angle Grinder and cutting Disc . Just rough out the curved shape. Leave the right extended portion long and wide at first as you can grind it off later to do the correct fit to the Sear.

Then use the bench Grinder and start bringing the finger curve towards shape. Be sure to hold the trigger piece in Vise-Grips as you grind and carefully cut the shape towards the result you want.

After a small amount of grinding, dip the piece in water; do so at intervals as you shape the piece.

And at the last, use a round file, and the Dremel Moto-Tool with a sanding cylinder to finalize and smooth the curvature to fit the finger.
This can be very nicely done, smoothed and curved to fit the finger perfectly!

The Sear

The Sear is easy to make... Just cut a 1 3/4" length of 1/4" thick steel, angling the Angle Grinder cutting disc to rough curve it to sort of match to the arc of hammer movement as you cut in from each end. Be sure to do a good V opening, leaving a good finger excess protruding below. And then grind the top carefully as you check the fit with the hammer location and the receiver space.

Check its match by comparing it on the receiver panel below the hammer and carefully grind it to a nice match on top to match to the hammer curve as it rotates. Do not finish the bottom yet.

Cut in the step at the left end of the Sear carefully, and smooth it as well. Especially the vertical part that will slide down to release the hammer. Drill the pivot point at the right end of the Sear with a 7/64" bit.

Once done, it is time to check where to mount on the receiver. Lay the receiver and mechanism on your work table.

You should have marked the cock point on the receiver from your firing test back in Chapter 6.

Temporarily remove the hammer spring from the receiver. Push the hammer back and mark the lower edge of the hammer onto a piece of tape attached to the receiver. This will give you a reference point for mounting the Sear.

Lay the Sear on the receiver and check that it is aligned to just clear the hammer as it swings back towards the firing pin, and also check that the notch step will be correct for catching the hammer at the cock position. (It is okay if the hammer lightly rubs the Sear piece during the cocking motion.)

Once you are sure of your location, mark the Sear pivot point onto the receiver with an awl or pen. Drill the receiver with the 7/64" bit.

Push a #6 Nail up through the hole from the rear of the receiver panel.

Recheck your mounting to be sure it is correct to lock the hammer, yet allow unrestricted rotation of the hammer. Once sure you can tack weld the head of the nail on the back of the receiver panel (outside of the gun.) Be sure the nail is vertical to the panel. Make a couple of spacers or temporarily use washers to slide over the nail pivot pin and slip the Sear onto the pivot.

Rotate the hammer back and check that movement is unrestricted and that the sear step can rotate upward to lock the hammer back. If everything looks good, it is time to set up the trigger and the rest of the trigger-Sear mechanism.

Later on, all of the parts will be surface hardened to give a nice smoothness and tough wear surface to each part.

Mounting The Trigger

If the trigger is finished to shape, (except for the Sear finger) set it onto the receiver below and to the left of the Sear. Keep in mind, the front upper extension of the trigger is to fit into the "V" of the Sear to pull it down.. Take care and do the fit up to check the fitting and imagine your handle later on the gun, and how your hand would match to the trigger. Determine the trigger finger extension angle needed to operate the Sear, and cut and grind it towards a fit for your setup. Grind any adjustment needed on the two parts for a smooth match for operation. Special care is needed for the trigger extension piece that will fit into the "V" on the Sear and the "V" itself.. Smooth these areas with the Dremel Tool or a file to do a nice smooth fit. The trigger must be able to rotate freely and turn the Sear CCW where you mount it.

Decide on the spot and drill the trigger's pivot hole with a 7/64" drill bit.

Then set the trigger in place, and fit the trigger to how you would mount it keeping in mind how it operates the Sear. Mark the pivot point on the receiver.

Drill the receiver with the 7/64" drill bit. Run a #6 nail up through the the hole and check that the trigger pivots nicely and will operate the Sear for releasing the hammer.

If everything looks good, tack weld a #6 nail to the outside of the receiver so the trigger can mount nicely.

My setup ended up with hole locations as shown in Figure 44. These are only approximate locations as these measurements are set using my actual hammer and Sear fit up to work with my setup for cocking of the gun. Another build may be different.

Move the hammer back to the cock position and check that he step in the Sear would lock it in place. Then operate the trigger to observe that it would rotate the Sear clear to release the hammer.

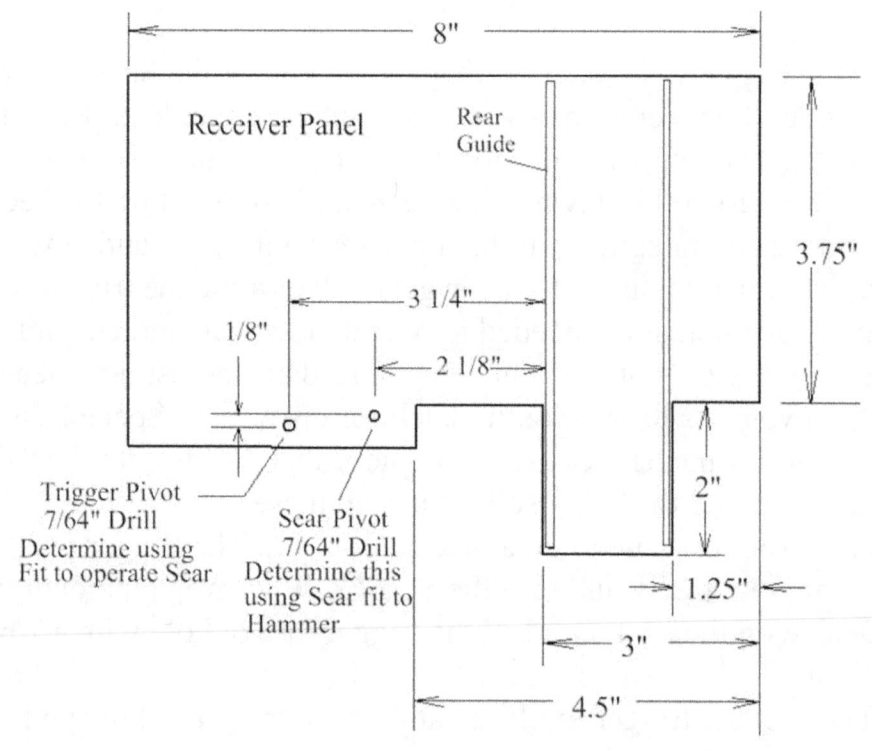

Fig. 44. My Approximate Trigger-Sear Pin Setup.

The trigger and Sear must be mounted centered between the

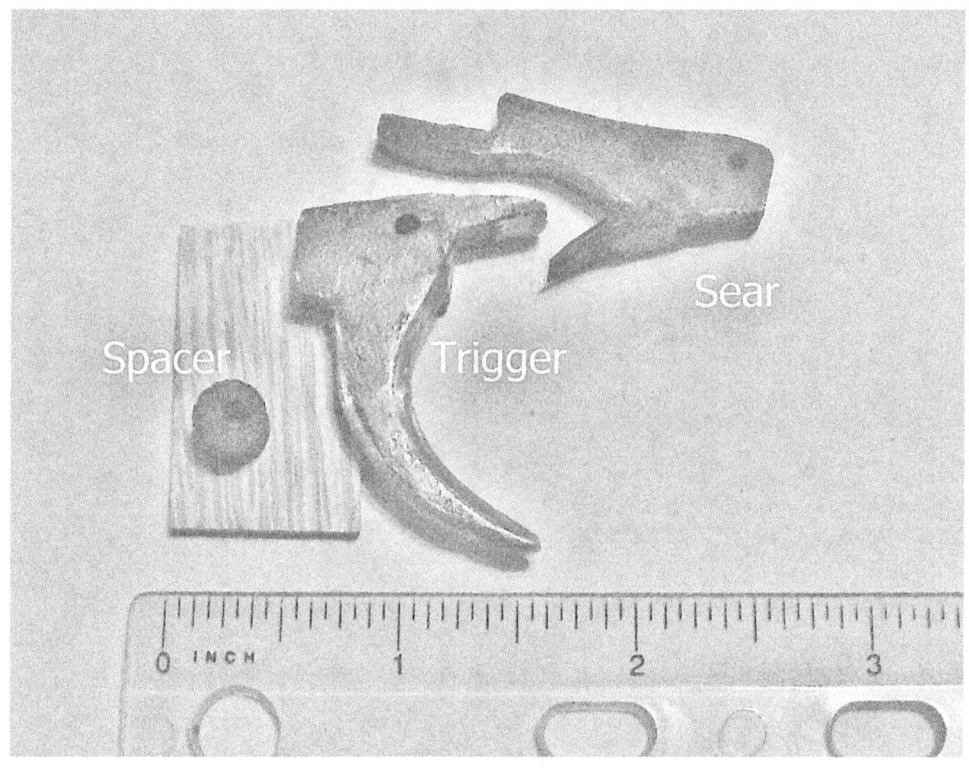

Fig. 45. View of the Trigger-Sear Parts.

receiver, so as to catch the hammer in a proper fashion. I made small wooden spacers to set this up. My actual parts are shown in Figure 45, above.

Making Spacers

Take a 1" length or smaller piece of 5/16" wood dowel, and mount it vertically in the drill press. Center a drill and drill through the piece with a 7/64" drill bit.

Then you can cut lengths from the dowel to make spacers to center the trigger and Sear between the receiver panels.

If all looks good; see Figure 46. Note the spacer also.

It is time to make and install the Positioner.

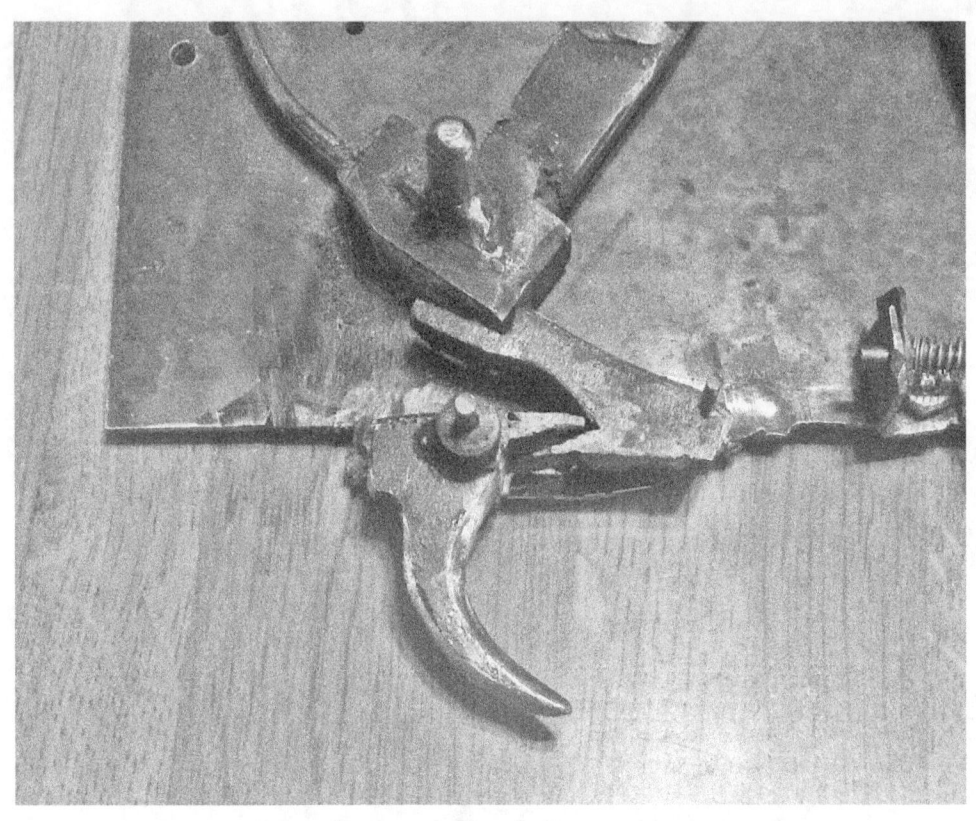

Fig. 46. Trigger-Sear Shown in Place.

Chapter 8

Construction and Install of Positioner

Making the Positioner

The Positioner is a simple part to make, ideally of 12 to 14 gauge steel. It will mount between the clevis on the hammer and is spring loaded to catch the rim of each round in the bar magazine and push each round into position for firing. Its final length will be close to that shown in Figure 47. Make it about 3" long, as you can true it to fit later. Angle the bottom tip as shown.

Note the nib at the left just below the pivot hole. This is for the spring which presses against the hammer clevis to apply force to the positioner as it catches each round during operation.

As the hammer is rotated to the cocked position, it must force the Positioner downward to move the bar magazine into place for each shot. The Positioner catches the rim of the round and forces the magazine down so the index ball grabs and locks just as the hammer also locks with the Sear. (The hammer catch.)

The lower 3/8" of the Positioner narrows and angles right a bit with a nice sharp tip which fits into the rear magazine guide slot and catches the rim of the .22 round.

The decision of where to pin the Positioner on the clevis is

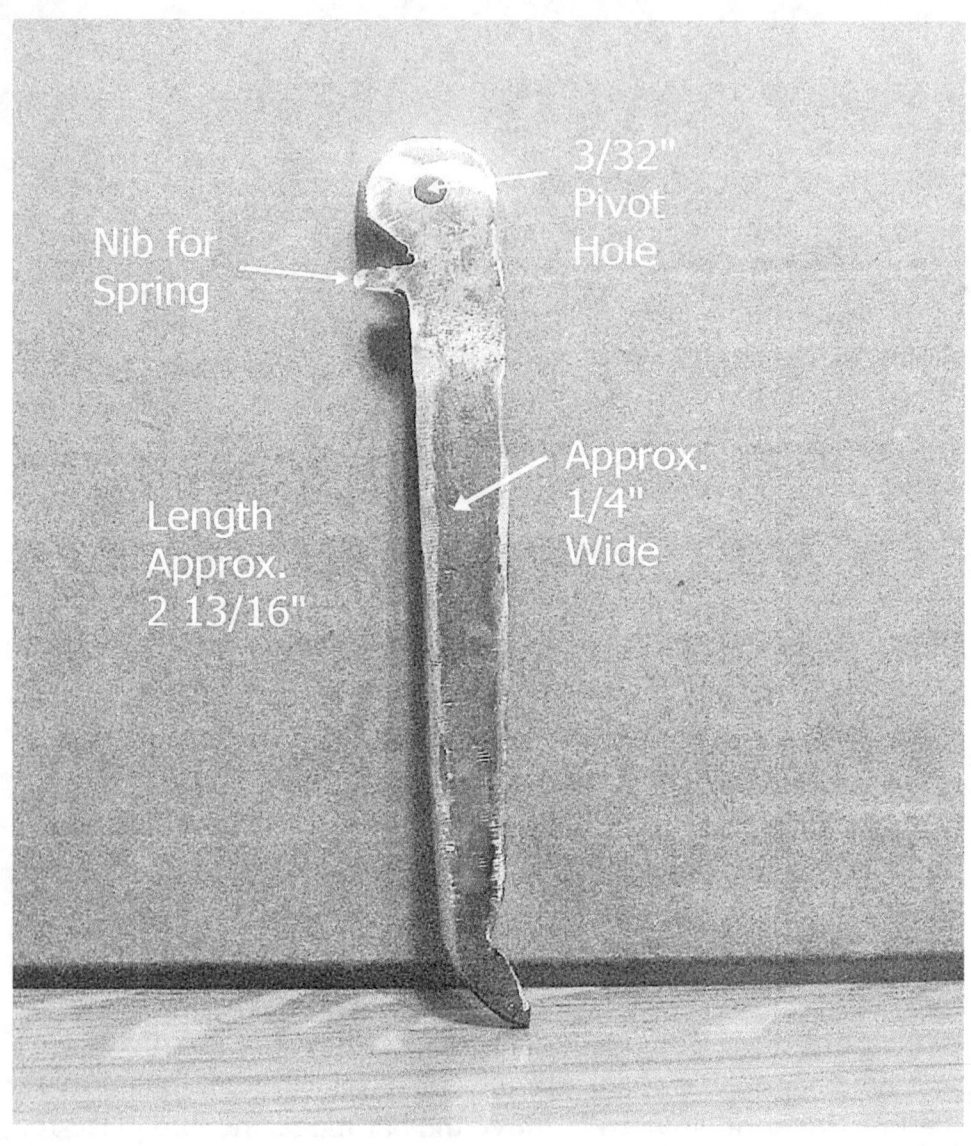

Fig. 47. Positioner Piece.

important as it has to work correctly as the hammer reaches the cocked position. This can be worked out with math or it can be done more easily with a piece of masking tape and some fit up. . I used math in a previous book, but the tape method is much quicker and easier.

Setting Up the Positioner With the Clevis

Remove the hammer spring. Place a narrow strip of masking tape on the clevis surface arm at the top of the hammer.

Place the hammer against the firing pin.

Hold the Positioner on top of the clevis, near the right end and centered top to bottom on the clevis. Make a mark on the tape through the Positioner pivot hole. Also place a mark at the lower tip of the Positioner near the rear magazine guide.

Now, rotate the hammer to the cock position and set the sear lock in place. Put the Positioner in place on the clevis again, aligning the pivot pin hole to the previous mark.

Now mark the tip again where it lies next to the rear magazine guide.

Measure the space between the two lines you marked for the tip of the positioner. *The total movement of the Positioner tip must be about 1/16" to 1/8" more than the spacing of the rounds in the magazine block. If larger, move the positioner in a bit on the clevis and do another mark. (If movement isn't enough, the clevis arm is too short.)*

More than likely though, the condition will be met with the clevis as I sized it.

1. Place some *fired* .22 casings into the bar magazine. Mount the bar magazine into the magazine slot and push it down to engage the *second* hole with the index ball. The lowest .22 cartridge and magazine hole will be aligned with the barrel opening.

2. With the hammer spring removed, cock the hammer back gently and set with the Sear step engaged.

3. Set the Positioner on top of the clevis, centered on the clevis, over the previously marked spot for the pivot pin.

Guide the lower Positioner tip to a point *just even with the top of the .22 rim at the bottom of the rear guide slot.* The top of the rim should be essentially even with the bottom of the guide slot in the rear magazine guide. With the top of the Positioner centered on the desired marked pivot point on the clevis arm, Mark the Positioner tip and file or carefully grind a bit off the tip so the tip just rests even with the rim inside the slot.

4. This is important so as to have caught the .22 rim and pushed the magazine downward for the first shot. Make sure to keep the top of the Positioner fairly centered on the clevis at the mark you made previously Verify the mark on the clevis tape through the hole in the top of the positioner. The Positioner should be at least 3/4 out towards the tip of the clevis with this spacing.

5. Hold the hammer handle, release the Sear from the hammer and lower the hammer to touch the firing pin.

6. Realign the Positioner to the mark on the tape. See if the tip of the Positioner is a slight bit above the .22 rim.

7. If the tip is a bit above, perhaps 1/16" to 1/8"you are set.

8. If it is below the rim, move the Positioner top slightly left or right on the clevis, then recheck as you leave the hammer touching the firing pin.

9. As you check the tip of the Positioner with the hammer touching the firing pin and then with the hammer cocked, you will shortly find a point on the clevis arm

that does meet the criteria.

10. The criteria is this: Hammer Un-cocked, and gently against the firing pin, the Positioner tip must be above the rim of the round, 1/16" or more. As the hammer is cocked, the tip must be essentially touching the rim of the lower round as the Sear locks the hammer and the round and magazine just seat with the index ball.

11. This is the point where the pivot hole of the Positioner should be on the the clevis, and where the clevis should be drilled for the pin to attach it.

Drill the clevis with a 3/32" bit catching both sides. Use a #4 Box Nail to make a pin. Cut its length slightly less than the space between the receiver panels.

Mount the Positioner on the clevis, putting the #4 Box Nail pin in with the head at the left side so it is held in by the receiver panel.

My spring pick for the Positioner was a silver colored 1" L x 3/16" Diam. spring from The Ace Hardware #1 Spring Assortment. Mount the spring between the nib on the Positioner and the nail nib at the inner part of the clevis. The spring pressure will keep the Positioner pushing against the .22 rims of the rounds in the bar magazine.

Once you set the spring in place, test your mechanism by this procedure:

1. Place the bar magazine into the firearm with expended .22 casings installed. Be certain the last partial hole is at the top of the magazine facing forward . Lock the magazine with the first hole aligned with the index ball.

2. Gently make sure the hammer is touching the firing pin.

The Positioner tip should be a bit above the rim of the round at the top of the rear magazine guide slot.

3. Now rotate the hammer to the rear as you watch the Positioner push the first round downward. As the hammer reaches the point of the Sear ready to lock it, the round should just align with the barrel and the index ball should pop in and lock the magazine in place for the first shot.

4. Repeat this action for each round, making sure to bring the hammer back to touch the firing pin, and checking that the Positioner has returned a bit above the round at the top of the rear magazine guide slot so that it will reliably feed the next round.

5. At this point you can re-install the hammer spring. Now do the cocking test as you *carefully use your finger to lock the Sear, and then BE SURE to hold the hammer to gently let it go forward to touch the firing pin.*

The Positioner could jump out of the slot during the shock of shooting later on so next make two small wooden blocks to glue on each side of the guide slot. Make sure they will restore the Positioner in place, but not bind the Positioner in any way.

See Figure 48 for my setup. This photo was actually taken later in the build but gives a good view of the Positioner guide blocks.

Fig. 48. View Showing Positioner Guides.

Chapter 9

Heat Treating the Parts

Surface Hardening

Next is the process of hardening and heat treating the parts. This is an easy process actually as the parts we are concerned with are all quite small.

The general procedure is to heat the desired parts to cherry red, then drop them instantly into oil.

I always do the process with used motor oil as it will have extra Carbon, due to engine waste during combustion. As the red hot part hits the motor oil, the metal surface absorbs carbon into the surface. The carbon and red hot metal combine in a thin layer of the surface, and the surface hardens significantly.

I usually do this twice, leaving the piece in the oil for a couple minutes. Then cleaning it off, reheating it again red hot and once again doing the oil treatment.

A bit of smoke comes off as you drop the red hot part into the can of oil. This is harmless but should be done in the open garage, and of course do not breathe the smoke. Along with the oil, keep a tin cover to place over the oil can just in case it should jump to a flame. Use a small can of oil, perhaps a tuna can. The parts are small. If it should flame up, simply put the tin panel over the can.

Tempering the Parts

There is a second part to the process as well, called Tempering; it is needed after hardening to lessen stresses in the metal and eliminate a brittle effect from hardening. The nice thing with this is you can do this in an ordinary oven!

This process requires heating the parts after hardening in an oven, at about 400 degrees for an hour, then letting them slowly cool.

First, the Hardening

Be Safe, use welding gloves, and safety glasses and have a fire extinguisher close, just in case.

The items needed are a Propane Torch kit, lighter, leather gloves, Safety Glasses, pliers and the small can of used oil. A metal surface to lay the parts on is good.

But the gas used is MAPP gas, which burns far hotter than Propane. It is ideal to heat the small parts red hot.

Use a pair of pliers to hold the parts as you heat them, be sure to use leather welding gloves, and safety glasses ! Have a fire extinguisher near.

1. Using a propane torch with MAPP gas, you will heat each part, or the desired surface to cherry red hot. Hold the part at the opposite end with a Vise Grip. There is an inner brighter portion to the flame. The tip of this inner flame is the hottest point, so try to keep it moving on the area to be treated. Heat the areas cherry red hot. Move the torch slowly back and forth if it is an area you are planning to harden. Once the area you wish to harden is cherry red, drop it immediately into the used oil.

2. Use a small can like a tuna can with the used motor oil. Used oil will have an extra dose of Carbon.

3. Continue this process with each part to be hardened. I hardened:
- the pivot pin for the the Positioner,
- the Positioner itself, both ends,
- The firing pin guide, L bracket,
- the rubbing parts of the Sear,
- The rubbing parts of the trigger, and
- The hammer spring guide and the clevis on the hammer.

Once you are done with the hardening process, remove from the oil and wipe the oil off each piece.

Notice how each piece has turned black where it was hardened, and there is a significant "slippery" feel to those areas too. It is a neat effect.

Time now to temper the parts to take any brittleness out.

Steps to Heat Treat, Tempering process.

Place the pieces on a Aluminum pie pan or a tin can and place in the oven set at 400 degrees. Allow a full hour in the oven at 400 degrees.

After an hour in the oven at 400 degrees, turn off the oven but leave the pieces in as the oven slowly cools.

Once the oven is cool, remove the pieces. This will complete the tempering process. The internal parts are finished, hardened and tempered.

See Figure 49. These are all the parts after the process.

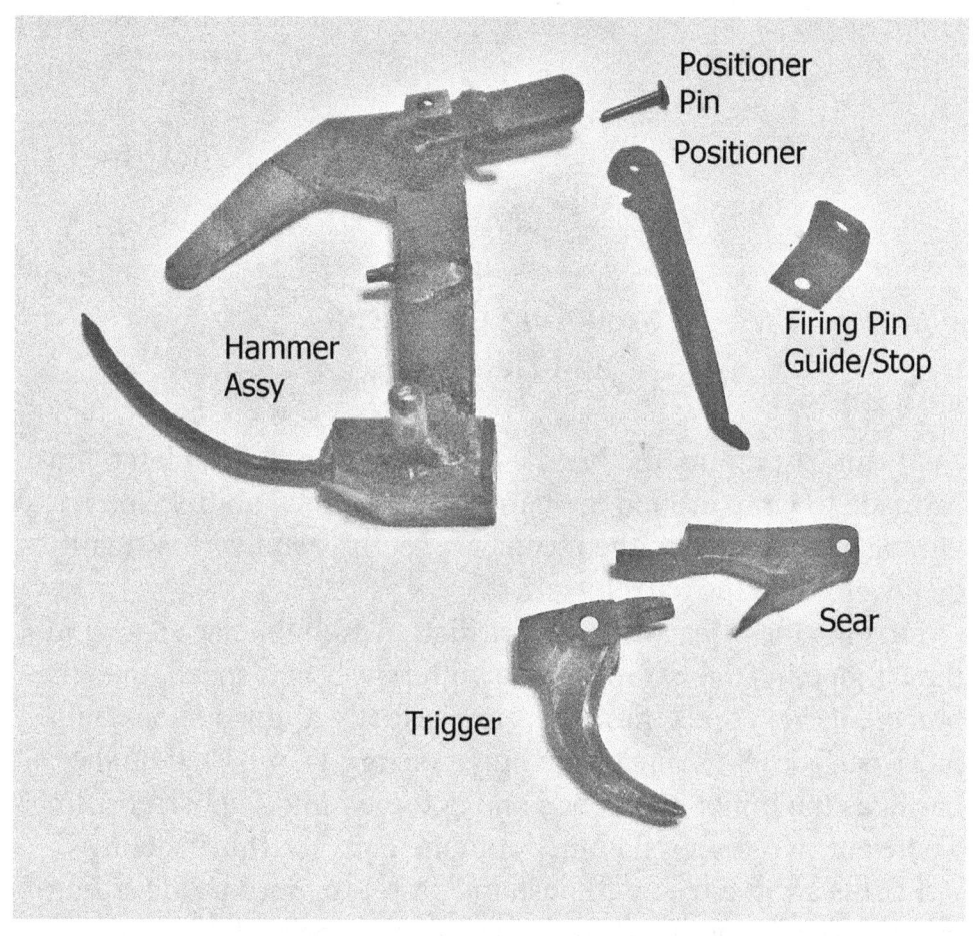

Fig. 49. Hardened, Tempered Parts.

Chapter 10

Mounting the Barrel and Hand Grip

It is time to mount the barrel. In order to do that, a clamp must be made to wrap around the barrel, with a pinch ability, and it will also be welded to the receiver assembly, and with a small tack weld onto the barrel itself.

Imagine the effect when a cartridge fires in the bar magazine-- the bullet exits the bar magazine with power and speed, then it hits the barrel cone – probably not perfectly aligned, close but NOT perfect. So some of the bullet energy is imparted on the barrel as the bullet is grabbed and is forced into the barrel rifling. There are two forces slammed into the barrel at this instant.

First is a force trying to push the barrel forward as the rifling grips the lead, and then a twisting force as the bullet is grabbed by the the twist rifling in the barrel. These are not paltry values. This is why rifles and other weapons have barrels either threaded in or pinned.

In this project, I made a round clamp of 12 gauge steel. It has tabs on the end which can clamp it tight around the barrel. On one side is also a tab, which can be welded to the receiver with a spacer. I wanted a very solid mounting and with the welding attachments, it seems more than adequate.

See Figure 50 for the clamp design. This steel strap must be bent into a loop to clamp the barrel to the receiver. The final hole to lock the barrel is to be drilled last to provide a tight lock as the

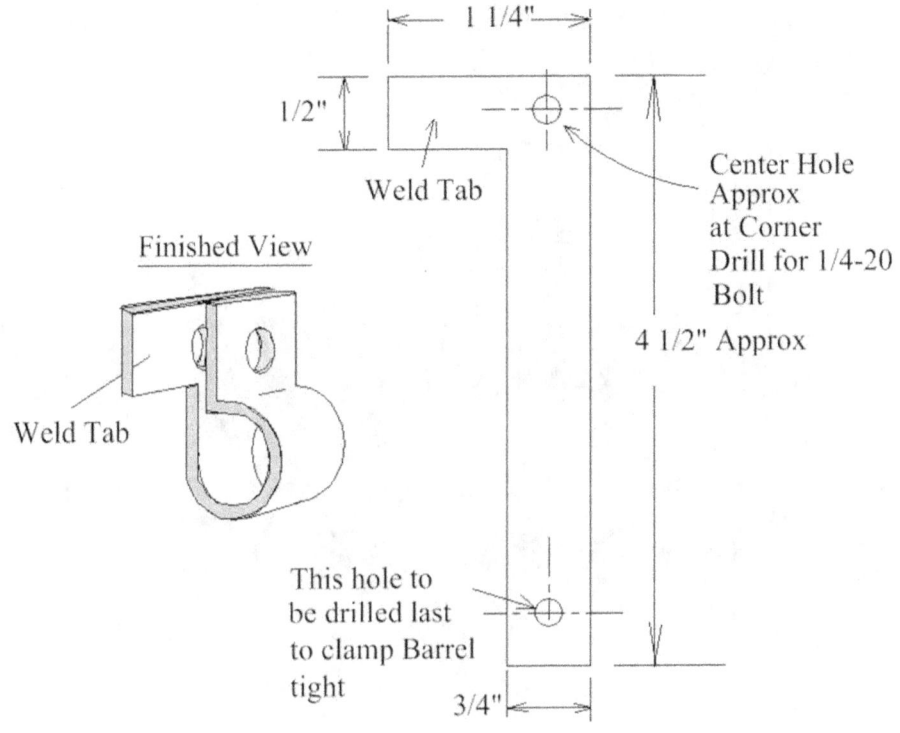

Fig. 50. Barrel Strap.

barrel is centered and mounted to the receiver. The weld tab allows the strap to be welded securely to the back (Left) receiver. It can be bent in a way to do this possibly with a spacer or piece to allow the barrel mounting to be very true .

If needed during bending of the strap, the propane torch and MAPP gas can be used to heat the areas as they are bent to a curve to fit around the barrel. The end is flattened to give a clamp surface ahead of the weld tab.

Once the part is fairly accurate to grip the barrel, it is slid over the rear of the barrel and slid forward.

It is a good idea to have the clamp on the barrel as you do the weld, so as to make sure the barrel is truly aligned as it is

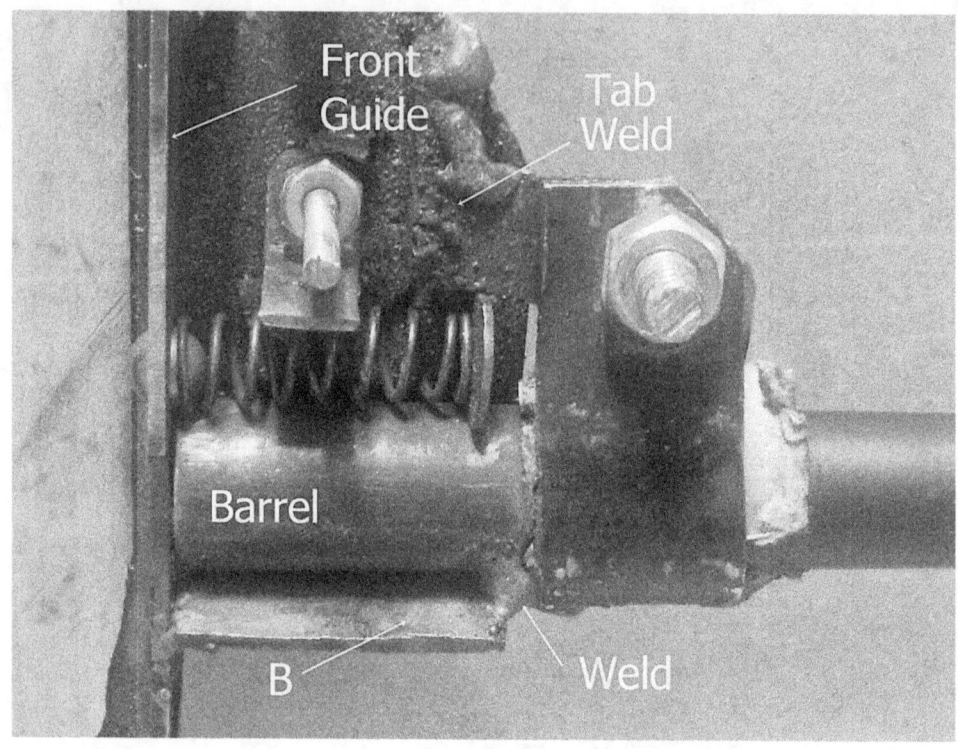

Fig. 51. View of Clamp Welded.

mounted onto the receiver. Push the rear of the barrel tight to the front guide. Then as the barrel is shimmed or clamped somehow to hold it true, and using a piece of wood, cardboard or metal to shield it from weld spatter, weld the weld tab onto the receiver using some shimming as well if necessary. See Figure 51. The photo shows a cardboard shim clamped around the barrel, also, but this was added later. The weld strap is shimmed and welded to the left receiver, and a small tack at the bottom attaches the barrel to the bottom piece, "B" which is welded to the receiver. This last part is done while securely clamping the barrel true and aligned to the receiver.

Figure 52 shows the assembly thus far. There are some spacers also welded on to mount the right receiver panel.

Figure 53 shows the inner parts more closely.

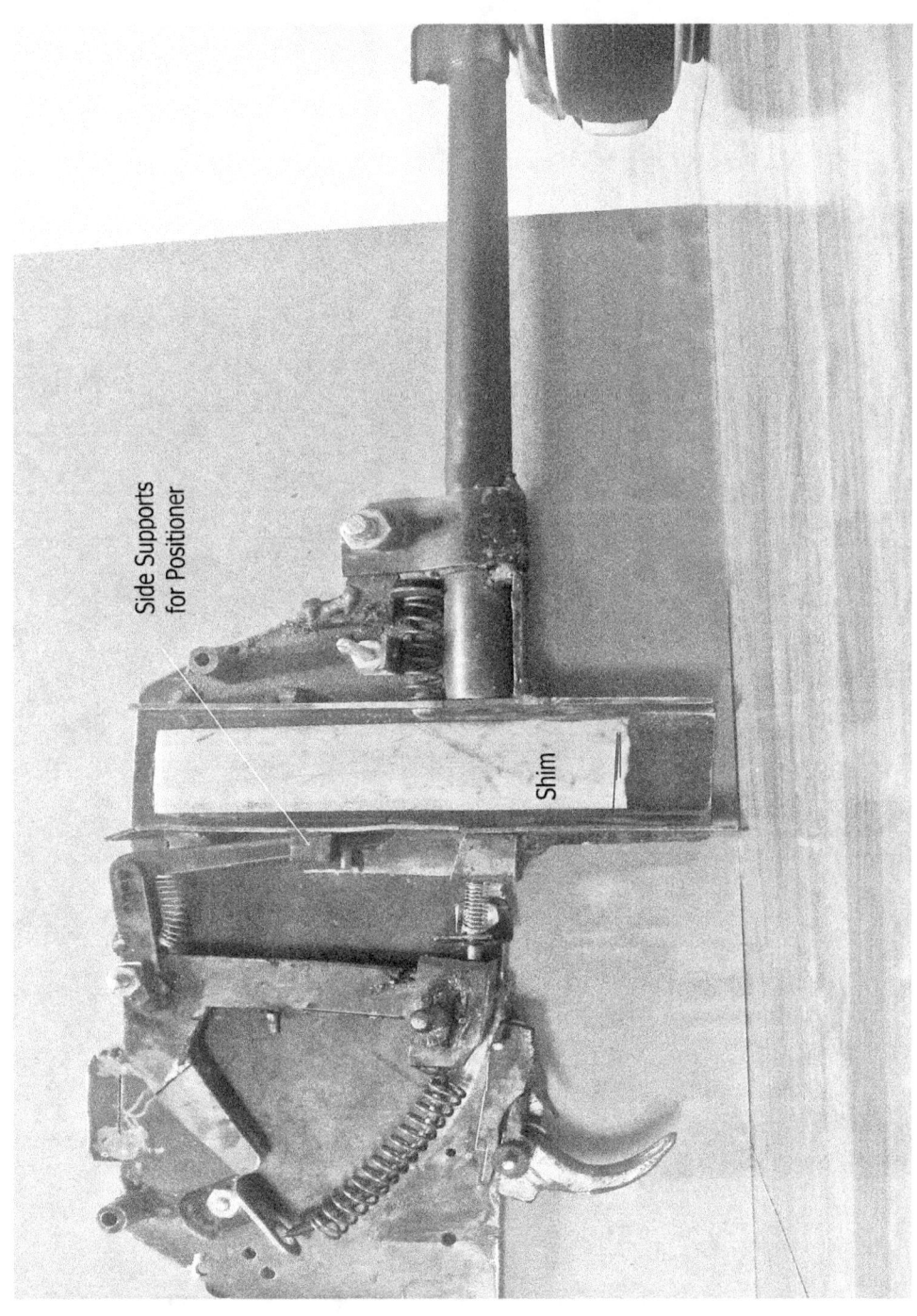

Fig. 52. Gun so Far.

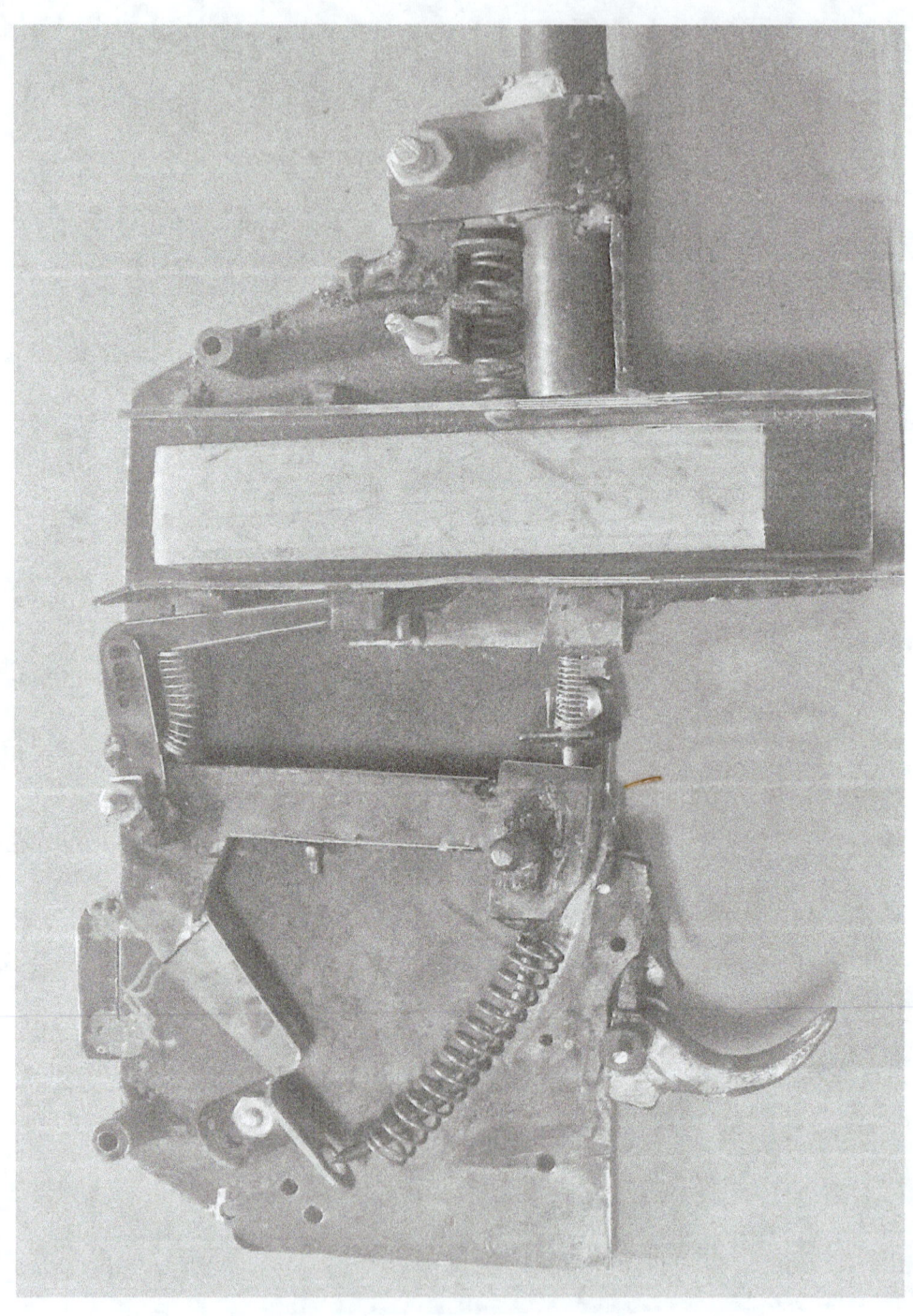

Fig. 53. Inner Parts View.

Testing the Barrel alignment

1. Get your .22 rifle cleaning rod. Remove the hammer spring.

2. Place **fired .22 casings** into the bar magazine. Load the bar magazine into the magazine slot, until you feel it lock the first hole with the index ball.

3. Touch the hammer to the firing pin, then cock the hammer as you check the Positioner is ratcheting the magazine down correctly. As the hammer reaches cock position, push the sear up, also note the index ball should lock into the second hole in the bar magazine at the same time.

4. Measure your cleaning rod on top of the barrel so it could reaches partway into the bar magazine. Place a piece of tape on the cleaning rod to mark that depth.

5. Insert the cleaning rod into the barrel and check that it does go all the way down the barrel and also *freely into the bar magazine. If so, you know the alignment is fine.*

6. Release the hammer and bring it forward to touch the firing pin, the Positioner should now engage the next round at its rim. Then do a second cocking action and check that the second round moves down and the magazine locks with the index ball. And now the second hole should be aligned with the barrel. *Check again that the cleaning rod goes freely into the magazine hole.*

7. Repeat this same check for each round in the magazine. If all are good, the fitting of the barrel and bar magazine are confirmed!

Spacers

Although I had already done some spacers for mounting the right receiver panel onto the gun, those can be done now if not yet done. Select some open areas where there will be no interference with any moving parts. Also avoid where the hand grip will go. The hand grip will be one spacer too..

I used 5/16" steel rod, cut pieces carefully to length to match spacing between the receivers (This is same as Barrel width at the breech...), then drilled through them on the drill press with a 9/64" bit. These are then welded at suitable points on the receiver. Run a bit once more through them after mounting.

Making the Hand Grip

I used various scrap pieces for my hand grip. But that was just because I had all sort of scraps. See Figure 54 which shows mine. Various style could be done based on a person's preference. The outer pieces are rounded to fit my hand.

The upper portion that will be between the receiver panels should be cut in from each side to leave a thickness matching to the space between the receiver panels. The area between the receiver panels to allow for the hand grip must be checked to determine where the inset cuts should be, and then the table saw or handsaw can be used to cut out the inset portion of the grip.

As this is done the piece can be checked until the desired fit is achieved. Figure 55 shows my general size dimensions.

As the mounting of the hand grip is determined, holes can be drilled in the receiver to mount the grip. During this work, you can visualize a spot to slip a piece of Pallet Banding steel into a

Fig. 54. Hand Grip.

thin cut in the handle to act as the spring for the Sear. This band can be seen on my handle in Figure 54. Make a thin cut into the front side of the grip as shown and embed a piece of pallet strap into the cut. Bend the right end upward to press against the left stepped portion of the Sear to provide the lock on cocking.

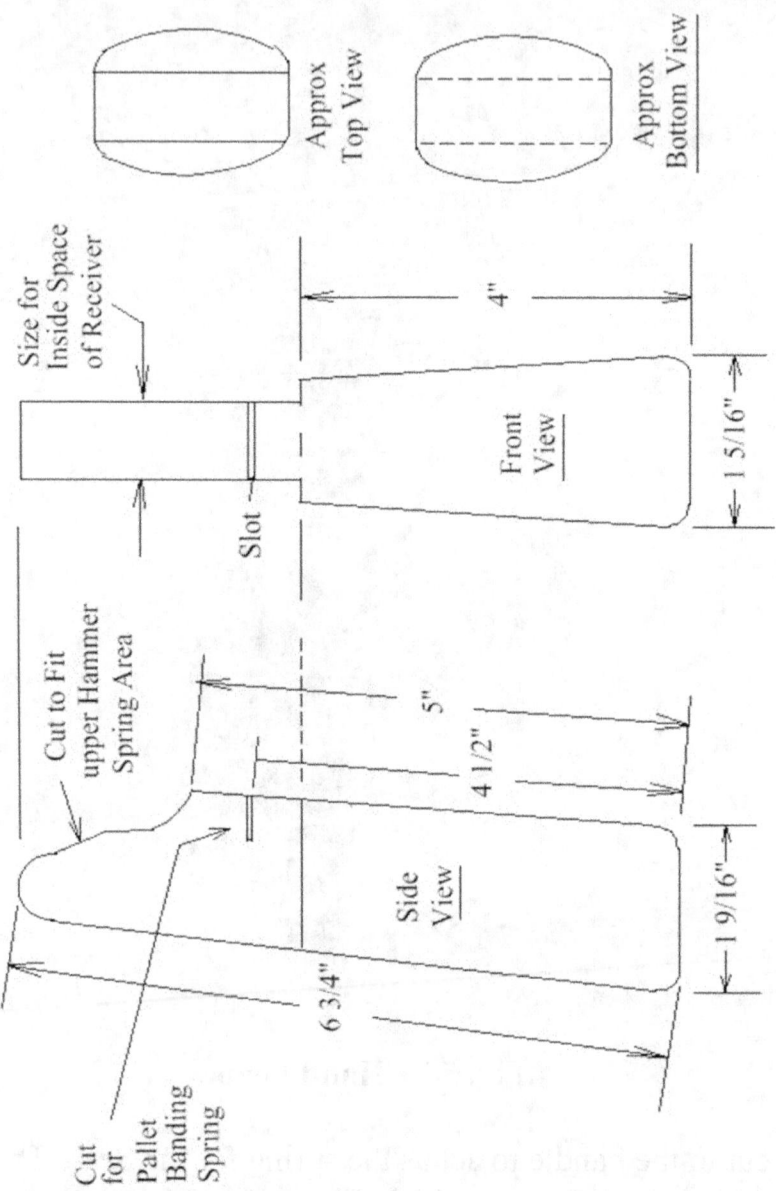

Fig. 55 My Hand Grip Build.

And Figure 56 shows the gun, now with handle mounted.

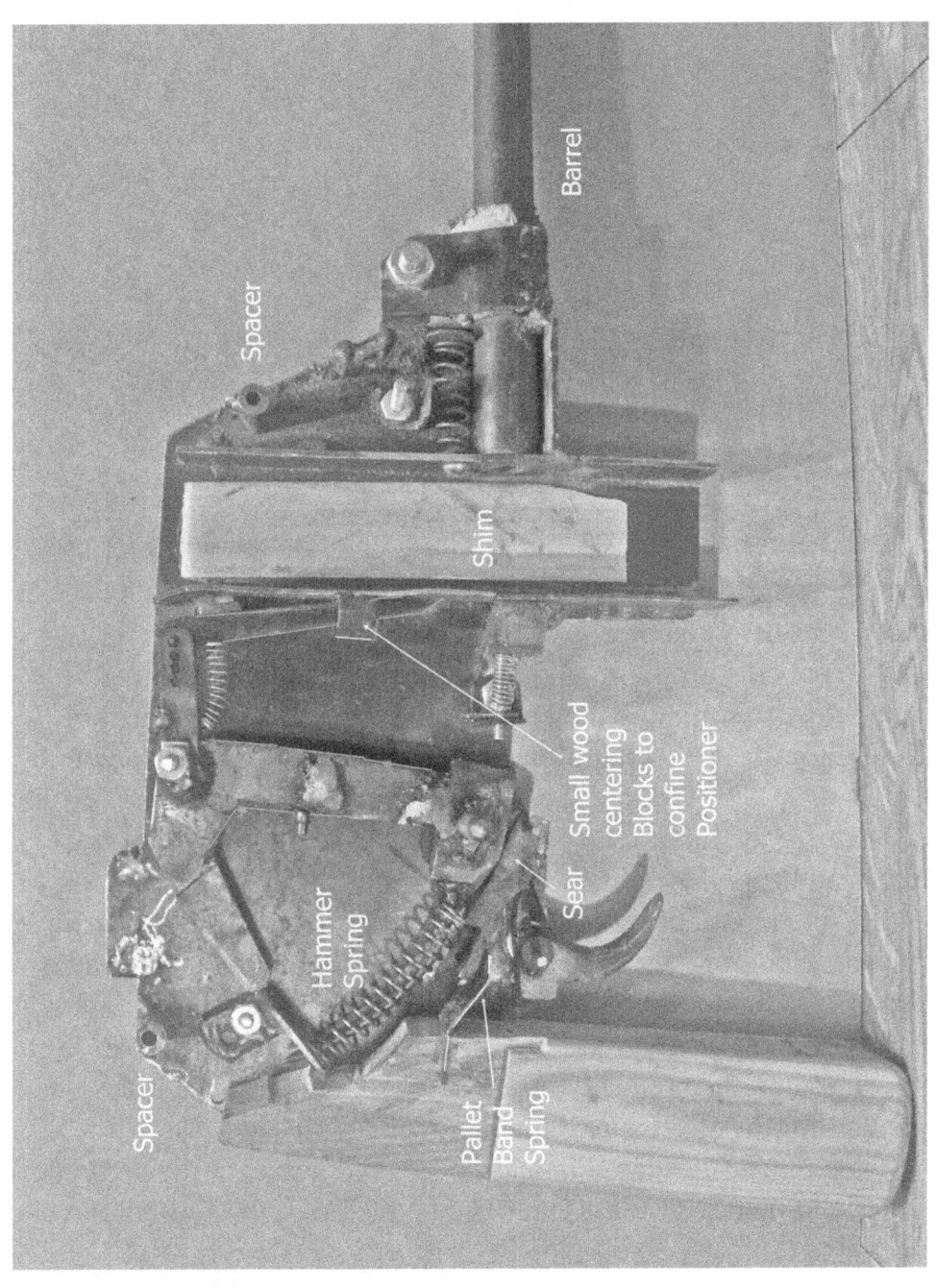

Fig. 56. Receiver Plus Hand Grip.

Chapter 11

Finishing Up and Mounting the Right Receiver Panel

All that remains now is the right receiver panel, a trigger guard and some sights. The right panel was also cut at the same time as the left back in Chapter 4.

The panel is pretty much solid, except an arced slot must be made to accommodate the cocking handle. And holes must be drilled for the spacers and the trigger, Sear and hammer pivot pins..

Remove the hammer assembly first, and its screw pivot to allow marking the right receiver panel for the spacer mounting.

Use the Dremel tool with a cutter disk to cut off the two nail pins on the left receiver panel for the trigger and Sear -cut them about 1/4" longer than the space between the receivers.

Lay the left receiver assembly on top of the right receiver panel; match them closely aligned. Mark and drill the holes needed for the spacers and the hammer pivot.

Then place a double thickness of masking tape in the area of the trigger and Sear pivot nails on the inside of the right receiver panel.

Use 6-32 screws through the left receiver spacers and the hammer pivot hole, to start guiding the two panels together as they will be when assembled.

As you do so, carefully keeping the alignment, push the right panel hard against the trigger and Sear pins to indent the masking tape and mark for drilling those two holes. Drill those holes with a 7/64" drill bit .

Now gently check the fit of the two together.

You can use 6-32 screws through the guides (later welding them onto the left receiver) to guide the receiver down. Check that both the front and rear magazine guides are even, that the receiver lies nice and flat on the guides, on the hand grip and each of the spacer mountings. File or grind any areas that are too high, until the right receiver panel fits nice and flat on the left assembly, over the spacer screws, the hammer pivot screw and the trigger and Sear pivot pins. If the panels lie nicely together, it is time to do the cocking slot.

You can also drill a hole in the right receiver panel over the hand grip for an equivalent mount point. Make sure as you look between the two panels that all surfaces match together pretty smoothly.

Later on you will tack weld the screw heads on one side so they will be mounted permanently for quick opening of the mechanism.

Making the Cocking Slot

Remove the right receiver panel. Remount the Hammer and Positioner over its pivot screw. No need for the Hammer spring right now. Place the hammer at the firing pin and very carefully measure it's location from the bottom of the left receiver and the rear magazine guide. Put a tape on the right receiver to mark where this point aligns to the rear magazine guide so you have two reference points that match to the left receiver panel.

Mark the spot on the right receiver that aligns with the center of the cocking handle. Drill carefully with a 9/64" bit at the center and align the receiver panels to check if the cocking handle seems aligned. If it looks close, re drill with a 5/16" bit. This should allow the cocking handle to go through the right panel, and the panel can lay flat with the various spacer screws and pivot pins in their proper alignment.. Check that this is the case.

   Place a small piece of paper aligned at the bottom of the receiver panel and large enough to trace the hammer movement, as it cocks, onto the left receiver panel aligned at the bottom of the receiver.   Rotate the hammer back to cock as you trace the curve onto the paper. Move the hammer back to the firing pin position.

   Now, align a mark about the center of the cocking handle . Remove your paper.

Redo your arc at about the center of the path the handle will go through in cocking the gun. Be sure that the hammer handle hole allows clearance for a full strike on the firing pin when forward, as this is when the Positioner must return above the next round ready for the next shot.

   Draw the arc on your right receiver in the location to match the center of the hammer rotation back to cock. Drill a series of 9/64" holes aligning each other along the arc about 1/4" apart or so.

   If a 1/4" cocking handle was used, drill each initial hole with a 5/16" bit to give clearance for a 1/4" handle. If a 5/16" cocking handle was used, drill with a 3/8" bit to open up the slot.

   Use the Dremel with a cutting disk or grinding tip or a file to help clean up the slot and curve it neatly to accommodate the cocking handle freely from firing pin to full cock position.

   If the job worked, the receiver should have a look like Figure 57 .

   A few small additions basically will complete the pistol. The right receiver panel should have an inner piece added to help bolster the back magazine guide on the right side, and a flange seal should be welded around the front and rear of the magazine

Fig. 57. Slot for Cocking Handle.

casing at the bottom of the pistol on the right receiver. The lower portion of the magazine guide serves as a front handle for the pistol, and having the additional seal of metal is a good safety item.

Also make the right shim for the magazine area to be sure the bar magazine feeds smoothly through without side slop. I made mine of wood.

Trigger Guard

A trigger guard can be added now to the bottom of the pistol receiver, and also will be fastened into the hand grip. Cut a strap 5/8" wide and 6" long of 14 to 16 gauge steel for this purpose.

Drill a 9/64" hole 5/16" from one end, centered, then bend an approximate 1/2" tab at this same end at an angle to attach to the hand grip. Hold the piece in place and visualize the curve you want to go past the front of the trigger; then bend the steel strap

into the curve. Using the torch with MAPP gas to heat the metal will make bending easier. At the upper right you must visualize flattening the strap to match to the bottom of the receiver. This right end of the strap can be welded to the left receiver in the area of the breech block you attached to the rear magazine guide in Chapter 5. Be sure to only weld it to the left receiver.
The left end of the trigger guard can be mounted to the hand grip.

Sights

The sights must be welded onto the left receiver panel once the pistol has been shot, the back sight will just be a "V" cut metal piece. Take care to bend a double "L" like in Figure 58 for the rear sight.

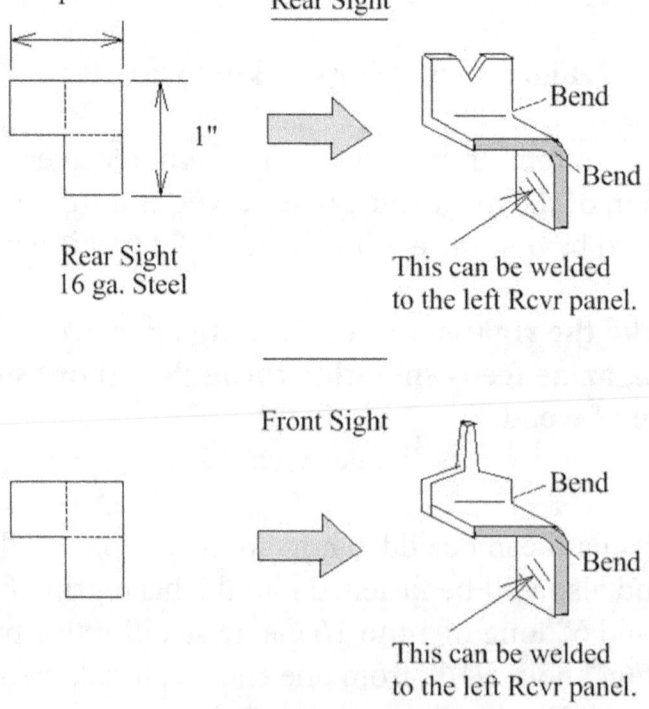

Fig. 58. Front and rear sight.

Fig. 59. Gun with Trigger Guard.

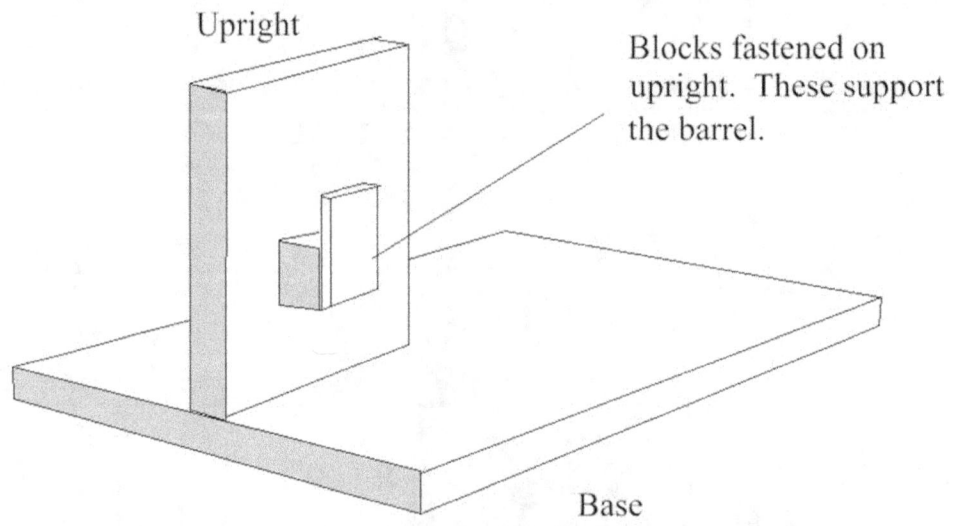

Fig. 60. Test Firing Jig.

The rear sight can be welded onto the left receiver piece at the top of the receiver, but leave the front sight for after the test firing.

Shooting tests

If you are finished with these last items, it is time to test fire the pistol. I recommend making a wooden base and clamp setup arrangement, to set the pistol securely anchored on the ground and allow a pull string for firing as you stand well back.

Look at Figure 60, which shows a simple test setup. The little barrel support block is mounted onto the upright at the height to support the barrel with the gun sitting fairly level.

The pistol can be set upright with the barrel in the notched block, and sandbags or some sort of weight can wedge to hold the pistol and base securely. Aim the pistol at a tree trunk, dirt bank or solid suitable backstop. A pull cord cord can be tied to the trigger, and passed back behind 30 or 40 feet or so to the rear.

1. Then a loaded magazine bar can be inserted. And pushed down to lock at the first index hole.

2. **Use only normal .22 ammunition, not hot types. I suggest Peters, Federal or Remington, NOT CCI, which might be a bit hotter rounds.**

3. Cock the pistol listening and watching that the magazine moves down and locks into the index position for the first round to be aligned with the barrel. You should clearly hear the click of the lock. (Also, you should have already tested this carefully with fired .22 rounds back several chapters.)

4. With the pistol cocked and ready to shoot, go back 30 or 40 feet to your pull string area. Pull the cord to fire a round.

5. If all seems good, and the pistol fired, proceed to the gun, check it over and re-cock the hammer noting that the next round locks into place correctly.

6. Proceed to the safe area and once again pull the cord. If firing appears normal, proceed with the test for each round until the pistol is empty.

7. After the pistol is empty, examine the bar magazine, and the gun carefully. Check for any cracks, bent magazine

guide, displacement or change in parts, the shims are still fine, welds, etc.-- anything at all. If everything appears normal it would be good to continue testing through at least 50 to 100 rounds.

8. When you finish, take the pistol apart enough to really inspect it inside. If everything still looks fine, it is a builder's decision as to further test or firing it from hands-on.

This is an expedient firearm, an experimental pistol, to show the capability of building a real firearm from scrap steel in a garage. After some more shooting and testing as you gain confidence, you can start to determine the sight setup, experiment first, then determine the mount and after shooting and perhaps marking it, grind the front sight carefully true as you install it. I hope to set mine for a 40 to 60 foot range with some accuracy.

This construction will involve doing windage and elevation considerations, so the height and centering of the front sight is critical . My feeling is to judge first with no front sight, but observing how far up or down and left and right the general area should be, perhaps taping a cardboard piece on the front to get close, then welding a suitable metal piece on the left receiver. The sight drawing in Figure 58 is only a general idea, the real front sight maybe end up taller or even adjustable if needed. With the front magazine as a forward grip and the 8" barrel, I expect it to be fairly accurate.

Remove Extra Metal Areas

You will see there are excess metal areas on the pistol, and probably noticed that I have cut off areas on the front and the upper rear as the later photos showed. It is only to reduce weight and streamline the pistol a bit.

And also add welded on pieces on the lower magazine guides, as shown in Figure 61..

This completes the construction.

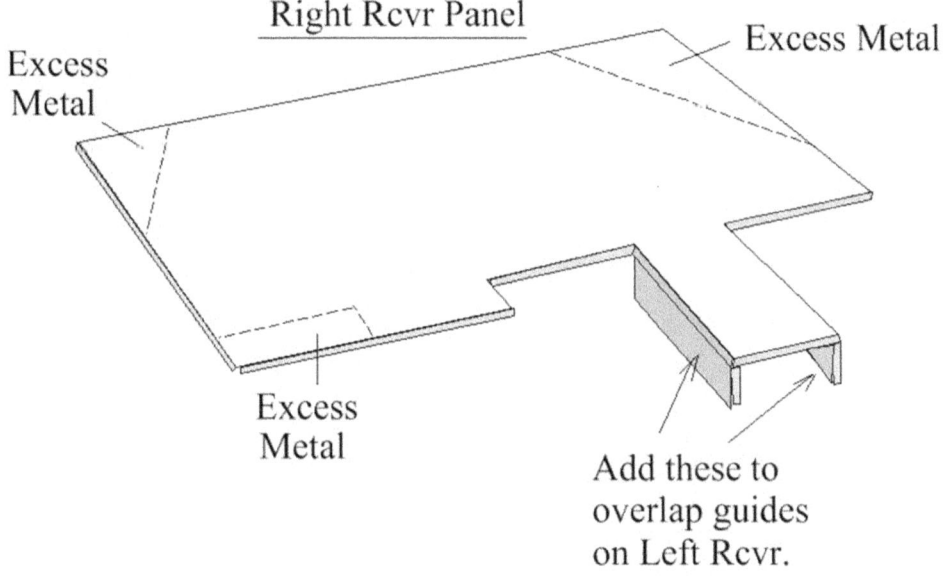

Fig. 61. Final left Receiver finishing.

Any hobbyist or person building this gun is doing so solely at their own risk, and it is their own responsibility toward safety, accuracy and workmanship. If you do so, be safe and do a good job.

112

www.ingramcontent.com/pod-product-compliance
Lightning Source LLC
Chambersburg PA
CBHW060516300426
44112CB00017B/2691